FOOD AND BEVERAGE CONSUMPTION AND HEALTH

CURRENT STATUS OF SUGARCANE RESEARCH IN INDIA

Food and Beverage Consumption and Health

Additional books in this series can be found on Nova's website under the Series tab.

Additional e-books in this series can be found on Nova's website under the e-book tab.

FOOD AND BEVERAGE CONSUMPTION AND HEALTH

CURRENT STATUS OF SUGARCANE RESEARCH IN INDIA

A. K. TIWARI
M. LAL
AND
A. K. SINGH
EDITORS

nova publishers
New York

Library of Congress Cataloging-in-Publication Data

ISBN: 978-1-63463-458-8

Library of Congress Control Number: 2014955181

Published by Nova Science Publishers, Inc. † New York

CONTENTS

PREFACE

Sugarcane is one of the most important crops commercially grown in about 115 tropical and sub tropical countries of the world. India is a major producer as well as consumer of sugar in the world and produced about 25MT of sugar from 360MT sugarcane in year 2011-13, contributing about 15 percent of the totals sugar production of the world. A quantum of sugar is produced from sugarcane, however, this crop faces a number of problems such as low cane productivity, biotic and abiotic stresses, high cost of cultivation, unavailability of seed cane of newly released varieties, post harvest losses and low sugar recovery.

In India sugarcane research was started in the beginning of 19th century. Since then a rapid advancement has been made in sugarcane cultivation by Indian researchers. The objective of this book is to provide comprehensive account of all the major achievements based on Indian workers in sugarcane research. The book is compilation of recent advancements made on sugarcane development, cultivation and on improvement in cane and sugar yield using conventional and biotechnological approaches by different agricultural scientist and researchers of India.

The book comprises a comprehensive discussion on research work done by the scientist/academician of India on different aspects of sugarcane development and cultivation such as Entomology, Pathology, Breeding, Physiology, Biotechnology, Seed production etc. The book will provide an up-to-date knowledge on sugarcane research being conducted in India to the graduate, post graduate students, research fellows, scientists/ professors involved in the field of sugarcane research and sugar industrialists in India and abroad.

This volume contains 11 chapters on various aspects of sugarcane development, cultivation, sugarcane agronomy, diseases, novel methods of seed multiplication, tissue culture, breeding and disease/pest management by distinguished sugarcane scientists from India. Three chapters focus specially on red rot disease which could be of immense importance in planning future strategies for disease management in India.

This publication of "Current status of sugarcane research in India" is purposeful in views of rapid research growth in India. Hopefully, it will prove to be modest and useful attempt in accelerating the pace of growth of researchers working on sugarcane in India.

Our sincere thanks are extended to all the authors for readily agreeing to contribute articles and for their timely co-operation in preparation of the manuscripts. We are also thankful to Dr G P Rao (Principal Scientist, Division of Plant Pathology, Indian Agriculture Research Institute, New Delhi, India) for his valuable help and encouragements. We are grateful to NOVA Publishing, USA for their determined efforts to publish the book on schedule.

We hope this book will serve as an important reference for students and scientists involved in sugarcane and related crops and stimulate research and extension work on burning issues of sugarcane.

INTRODUCTION

The book describes various major diseases of sugarcane seedlings and standing crops in Bihar. The occurrence of diseases, losses incurred, and their effective control measures have been illustrated.

Sugarcane micropropagation for rapid seed production has been enormously emphasized in recent years; however, adaptability of micropropagated planting material has not been up to the desired extent. Useful strategies have been suggested for production and management of quality seed of sugarcane through micropropagation. Utilization of tissue culture derived variation in sugarcane improvement and the developments made in the field of *in vitro* induced variation and its use in sugarcane improvement without disturbing the genetic constituent of a clone have also been described.

The book also includes an article on a simple and cost effective agricultural innovation called the 'Sustainable Sugarcane Initiative (SSI)' which can be applied to sugarcane cultivation using comparatively less inputs, seed, water, and fertilizers. They have advocated the use of bud chips instead of 3-bud setts as planting material and transplanting of the seedlings raised from bud chips with wider row spacing. Better plant stands and higher cane yield of 105.0 t/ha have been obtained using SSI technology as compared to the cane yield of traditional three bud setts planting, 89.0 t/ha.

Some weather parameters have been correlated with the incidence of Early Shoot Borers (*Chilo infuscatellus*) and Scale Insects (*Melanaspis glomerata*) in Sugarcane in the North Coastal Region of Andhra Pradesh, India. This study can be useful in the development of a decision support system to manage the ESB and scale insect.

Limitations such as complex genome, narrow genetic base, poor fertility, susceptibility to biotic and abiotic stresses and long duration to evolve elite cultivars impose challenges in sugarcane through conventional breeding methods. Application of biotechnological tools for sugarcane improvement has been discussed in detail.

A protocol for fermentative production of good quality vinegar from sugarcane has very nicely been illustrated which can be very useful to the distillery and sugarcane byproduct industries.

Abiotic stresses such as drought, salinity, water logging, and temperature extremes are the most important factors limiting cane productivity. The responses of sugarcane to these abiotic stresses and appropriate measures for sugarcane management under stress conditions have also been discussed.

A. K. Tiwari, M. Lal and A. K. Singh, Editors

In: Current Status of Sugarcane Research in India
Editors: A. K. Tiwari, M. Lal and A. K. Singh
ISBN: 978-1-63463-458-8

Chapter 1

IMPACT OF WEATHER PARAMETERS ON THE INCIDENCE OF EARLY SHOOT BORER (*CHILO INFUSCATELLUS*) AND SCALE INSECT (*MELANASPIS GLOMERATA*) IN SUGARCANE IN NORTH COASTAL REGION OF ANDHRA PRADESH, INDIA

B. Bhavani, B. Bapuji Rao and N. Venugopala Rao
Regional Agricultural Research Station, Anakapalle,
ANGRAU, Andhra Pradesh, India

ABSTRACT

In Andhra Pradesh, sugarcane crop is subjected to a limited range of about 15 insect pests, of which early shoot borer *(Chilo infuscatellus* Snellen) and scale insect (*Melanaspis glomerata* Green) are the regular serious pests causing yield loss (both of cane and sugar content) and making extensive replanting necessary in many parts of Andhra Pradesh in India. Research on the impact of weather parameters on the incidence of early shoot borer (ESB) and scale insect at different stages of the crop growth in sugarcane showed a strong association of ESB incidence with minimum temperature and morning relative humidity. Relatively warm (minimum temperature > 23.8°C) and dry nights (RH < 77%) favoured the incidence of ESB. The positive correlation between maximum temperature and incidence of the ESB is significant only during 45-60 days after planting. However, minimum temperature showed a strong association with the ESB incidence during major part of the crop season. The association between mean temperature and the ESB seems to be dominated by minimum temperature. Early morning relative humidity was found to profoundly influence the ESB infestation and high humidity reduced the infestation levels. High rainfall events with rain exceeding 50 mm/day are detrimental to ESB during the early stages of crop growth.

It is evident from the present results that when high maximum temperature and low relative humidity prevails, *C. infuscatellus* is active as shoot borer during May-June and drought conditions coupled with low rainfall enable early shoot borer to continue as internode borer in North Coastal Region of Andhra Pradesh under rainfed conditions. The activity of borer is reduced with the receipt of monsoon rains. Light showers and cloudy weather are detrimental for the multiplication of C. *infuscatellus*.

Mean temperature and relative humidity during 7-12 fortnights after planting (FAP) showed significant correlation with the infestation of scale insect, *M. glomerata*. As the mean temperature during 7-12 FAP increased, the scale insect incidence also increased. The association between these two was better expressed by a quadratic fit rather than a simple linear regression. Mean relative humidity (RH) during this period showed a significant negative correlation. Lowest incidence was noticed when the RH exceeded 80%, coinciding with high rainfall events. Low rainfall years leading to low RH during 7-12 FAP period favoured high incidence of scale insect on the sugarcane. Rainfall during tillering and early cane formation stage (1-12 FAP) was negatively correlated with the incidence of scale insect as high rainfall events might have washed away the pest. Rainfall of above 500 mm limited the incidence of scale to approximately 10% and less than 400 mm rainfall during 11-12 FAP led to high incidence.

Keywords: Sugarcane, *Chilo infuscatellus*, *Melanaspis glomerata*, correlation, regression, temperature, relative humidity, thermal time, growing degree days (GDD) and rainfall

INTRODUCTION

Sugarcane is one of the major industrial crops in India. It is cultivated under diverse agro-climatic conditions. Though India tops the world in the total area under sugarcane cultivation, the average yield per unit area is low. Sugarcane is grown over 2 lakh hectares in Andhra Pradesh whereas in north coastal zone of Andhra Pradesh state, sugarcane is grown in an area of 70,000 hectares and of which 80% is under rainfed conditions. The productivity of cane in north coastal zone is 60 tonnes per hectare as against the state average of 76-78 tonnes per hectare. This low yields under rainfed conditions could be attributed partly to moisture stress and partly to key insect pests like early shoot borer (*Chilo infuscatellus*) and scale insect (*Melanaspis glomerata*) damage. High day temperatures coupled with moderate relative humidity is more conducive for ESB multiplication [1] during early stage of the crop growth. In severe cases of infestation, the damage due to ESB could be as much as 90 per cent [2]. Though the early shoot borer generally attacks the shoot stage, it is also found to act as cane borer in Rajasthan [3, 4], West Bengal [5] and Andhra Pradesh [6]. This peculiar behaviour of the pest has a relationship with weather parameters. Avasthy et al., [7] reported that drought conditions and low rainfall enable shoot borer to continue as internode borer.

Early shoot borer infestation is high during the pre-monsoon period (April-June). Heavy borer infestation has been observed when high temperature prevails with low to moderate humidity. Borer activity decreases appreciably with the break of south-west monsoon. Infestation in malleable canes occurs in Andhra Pradesh, Orissa, Rajasthan and Tamil Nadu where temperatures raise appreciably during the post monsoon period (September - October) or in years when the rains cease earlier than normal. Attempts have been made in the past to find out range of maximum temperature favouring increase in population. Kalyanaraman et al., [8] reported that *C. infuscatellus* is essentially a pest of the pre-monsoon period. The moth of the borer being a nocturnal one, its activity is related to moonlight. Prasada Rao et al., [2] reported that the moth of the ESB being a nocturnal one, its activity is related to moonlight.

The scale insect, *Melanaspis glomerata* has adapted to a wide variety of climatic conditions. Its infestation has been noticed from the regions with moderate temperature and drought conditions of Madhya Pradesh, Maharashtra and Gujarat to waterlogged areas of eastern Uttar Pradesh and Bihar. Its incidence was also observed in areas with extreme temperatures and moderate humidity of western Uttar Pradesh and Delhi to moderately warm and humid conditions of Coastal Andhra Pradesh and Tamil Nadu. Dry conditions predispose the crop for scale insect activity. It is evident that soil moisture stress favourably influences scale insect population persists in the Coastal districts of Andhra Pradesh, where frequent irrigations are given in the pre-monsoon period and water logging is a common phenomenon during the monsoon months. Maximum population is found during July to October, when high temperature and humidity prevail. The scale insect survives on the setts though covered by soil and with the formation of internodes the infestation builds up. Infestation commences prior to the start of monsoon rains in early plantings and ratoons. Its build up is mainly from July onwards till harvest. A recent estimate by Ministry of Agriculture, Government of India noted that scale infestation reduced cane yield by 32.6% and sugar recovery by 1.5 - 2.5% [9].

Many studies were conducted by several research workers in the past to relate the incidence of ESB to planting time [2, 10]; varieties/genotypes [7, 11]; soil factors [2], sucrose content [12] and climatic factors like temperature [13] and humidity [14]. Several studies conducted earlier at Anakapalle indicated the role of weather parameters on the incidence of ESB [2, 15] but a lone parameter could not be identified in entirety accounting for the incidence of ESB. Several attempts were also made in the past to relate the incidence of ESB and scale insect to climatic factors like temperature [13, 16], humidity [14] and rainfall [10]. Studies conducted earlier at Anakapalle indicated the role of weather parameters on the incidence of ESB [15] and scale insect [17].

The research on the role of weather on the incidence of early shoot borer and scale insect (*M. glomerata)* in sugarcane ecosystem using a ten year data, the possibility of forecasting their population outbreaks, correlation and regression studies to find the relationship between weather parameters and insect pests using SPSS 16.0 software package are presented in this chapter.

Inter-Annual Variability

The mean incidence of the ESB was high in the early stages of crop (45 and 60 DAP) and gradually decreased beyond 60 DAP (Figure 1).

Sugarcane germinates by one month after planting and thereafter the shoots are exposed to the pest. As the moth takes a month's time to complete its life cycle, infestation is likely to commence from 40 DAP and may reach the peak during 45 to 60 DAP.

The ESB passes through nine generations in one calendar year at Anakapalle [15]. Thus, normally the crop is subjected to the infestation of ESB of 3-4 generations during the formative phase. Pest attack during this critical phase drastically reduces the crop yield [2].

There is a high inter annual as well as inter periodic variability in the incidences of the ESB and scale insect (Figure 2) which could be due to manifestation of several factors including weather.

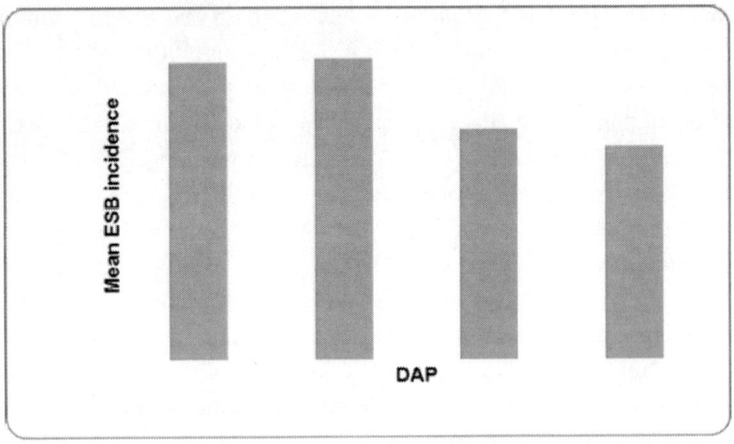

Figure 1. Mean per cent incidence of ESB at different stages of sugarcane.

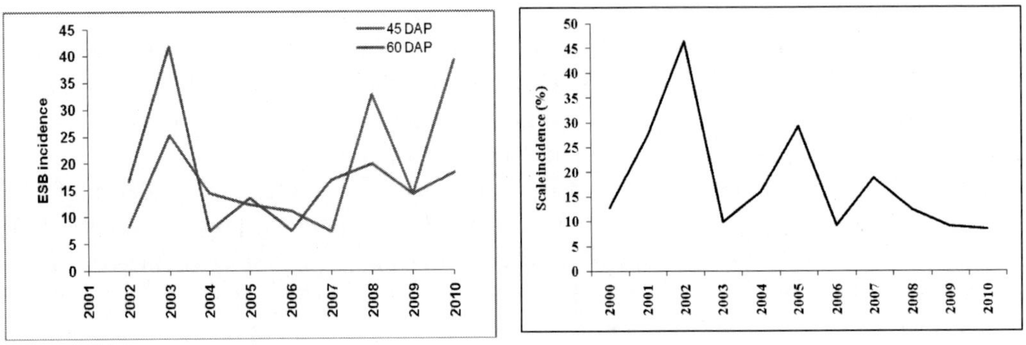

Figure 2. Inter-annual variability of ESB peak incidence and scale insect incidence.

The highest incidence of scale insect was recorded during 2002 followed by 2005 and lowest incidence during 2010. The year 2002 was a low rainfall (724 mm) year whereas 2010 was a high rainfall (1717 mm) year. However, the rainfall during 2005 was relatively high (1212 mm) during which the incidence of *M. glomerata* was also considerably high. It can be thus inferred that the incidences of *C. infuscatellus* and *M. glomerata* are affected to a large extent by prevailing weather conditions, as no pest control interventions were implemented.

Temperature and Relative Humidity

The weekly maximum temperature ranged between 33.3 to 39.6°C and minimum temperature between 17.6 to 29.5°C during the peak period of ESB incidence (45-60 DAP). Morning relative humidity (weekly mean) ranged between 80 to 93 per cent and afternoon relative humidity between 38 to 59 per cent during the corresponding period. The correlation between maximum temperature and incidence of the pest is significant only during 45-60 DAP (Table 1). Karla [18] reported that high temperature (35 to 38°C) and low relative humidity are favourable for the activity of shoot borer at Sriganga nagar whereas Tanwar and Bajpai [19] reported a significant positive correlation between maximum temperature with borer incidence.

Table 1. Peasson's correlation between weather parameters and ESB infestation at fortnightly intervals

Weather parameters	15-30 DAP	30-45 DAP	45-60 DAP	60-75 DAP	75-90 DAP	90-105 DAP	105-120 DAP
Max. Temperature (°C)	0.30	0.43	0.64*	0.25	0.24	0.20	-0.02
Min. Temperature (°C)	0.46	0.18	0.63*	0.63*	0.75*	0.73*	0.44
Mean temperature (°C)	0.44	0.28	0.67*	0.52	0.67*	0.60*	0.18
Daily temperature range (°C)	-0.42	0.05	-0.37	-0.63*	-0.45	-0.51	-0.41
Morning relative temperature (%)	-0.61*	-0.62*	-0.43	-0.48	-0.50	-0.62*	-0.13
Evening relative temperature (%)	0.50	0.07	0.23	-0.43	-0.03	-0.19	-0.11
Total rainfall (mm)	0.43	-0.02	-0.04	-0.27	-0.20	-0.19	-0.06
Total evaporation (mm)	0.26	0.29	0.55*	0.50	0.35	0.46	0.22
Hours of Bright sunshine (hrs/day)	0.18	0.22	0.40	-0.20	-0.03	-0.07	0.06

*Significant at 5% level; DAP: days after panting.

However, minimum temperature showed a strong positive association with the pest incidence during major part of the crop season. The association between mean temperature and the pest seems to be dominated by minimum temperature. The diurnal range of temperature behaved as that of maximum temperature. In the earlier studies also the conducive role of maximum temperature was noticed [11]. In a two year study on light trap catches of ESB moths, Rao and Babu (2004) found significant positive correlation between moth catches and maximum, minimum temperatures. The maximum atmospheric temperature during the oviposition period was found to be positively correlated with the total population and incidence of the borer [20]. High day temperature coupled with moderate relative humidity is conducive for the multiplication of early shoot borer [8].

The work of Krishnamurthy Rao [1] indicated that the borer multiplied profusely during summer months and preferred high temperature and low relative humidity. Varadarajan et al., [21] concluded that maximum temperatures of 35.6°C and 36.5°C with low relative humidities of 78.2 and 81.3 per cent during May and June respectively, were highly favourable for the large scale multiplication of the shoot borer, *C. infuscatellus*.

Hapase et al., [22] observed a significant positive correlation of temperature with borer infestation while relative humidity at a negative relationship. This is in confirmation with the laboratory observations of Pradhan and Bhatia [23]. It is thus evident that for borer multiplication, the most important abiotic factor is temperature.

High morning relative humidity in the present investigation was found to reduce the ESB infestation (Table 1). It is interesting to note the contradictory role of minimum temperature and relative humidity. For a better explanation of these abiotic factors one has to look into the insect behaviour. The moth of the insect is nocturnal and its movement and ovipositional activity is mainly during night time. High night time temperatures coupled with low humidity might have favoured the moth movement during nights. Conversely, cool nights coupled with humid weather must have curtailed the moth movement or larval hatching or both. It is evident from the data that when high maximum temperature and low relative humidity prevail, *C. infuscatellus* is active as shoot borer during May and June. In Andhra Pradesh, the probable reason for early shoot borer acting as intermodal borer is the practice of planting cane in June-July under rainfed conditions.

The crop planted during this period germinates by mid-August when the temperatures are fairly high. Availability of young shoots in younger crop during September – October, which in turn helps in the buildup of the pest population which infests the grown up crop as cane borer. The similar cases of infestation have been reported from some areas of Orissa, Madhya Pradesh and Maharashtra.

Among the various weather parameters affecting the infestation of scale insect, mean temperature and relative humidity during 7-12 FAP showed significant correlation. As the mean temperature during 7-12 FAP increased, the scale insect incidence also increased. The association between these two was better expressed by a quadratic fit rather than a simple linear regression (Figure 3). Mean relative humidity (RH) during this period showed a significant negative correlation (Figure 3). The association between these two measures was explicitly expressed by a quadratic fit. Lowest incidence was noticed when the RH exceeded 80%, coinciding with high rainfall events. Low rainfall years leading to low RH during 7-12 FAP period favoured high incidence of scale insect on the sugarcane. The 7-12 FAP coincides with the grand growth phase in sugarcane, which is the most important stage in the life cycle of this crop because the actual cane formation and elongation and thus yield build up takes place during this period. The pest attack during this phase drastically reduces cane yield. The present findings are in agreement with Krishnamurthy Rao [17] who reported that dry conditions predispose the crop for scale insect activity.

Rainfall

Rainfall during the 60-90 DAP was observed to reduce the ESB infestation though the association was not significant (Table 1). The activity of the ESB is reduced with the receipt of monsoon rains. Barring few occasions, the rainfall intensity in most of the years was below 50 mm per day, which might be the reason for the low correlation coefficients with rainfall. During the year 2010, a rainfall of 60.4 mm on 50 DAP and 88.6 mm on the subsequent day (51 DAP) resulted in decline of percent incidence of ESB from 9.68 to 0.99 by 60 DAP. Except this occasion, the influence of rainfall on ESB could not be established.

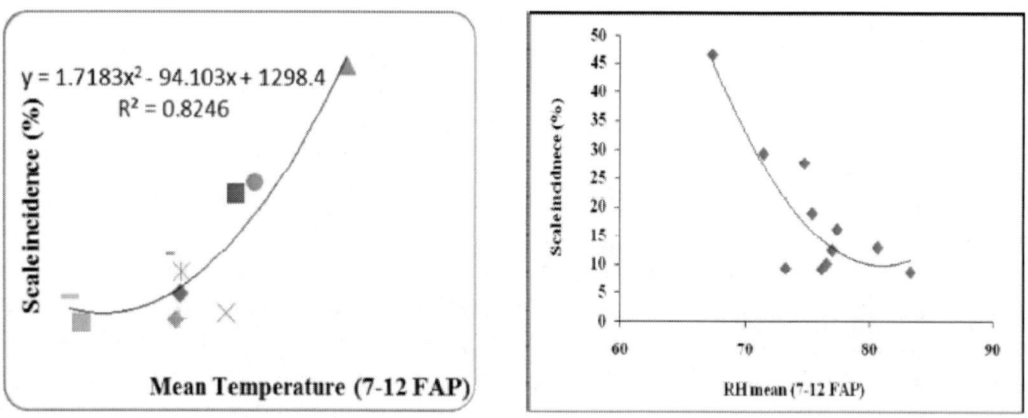

Figure 3. Relation between mean temperature (°C), relative humidity during 7-12 FAP period and scale insect incidence.

Sithananthan et al., [10] reported that lesser rainfall appear to be favourable for the borer multiplication whereas Avasthy et al., [7] reported that drought conditions and low rainfall enable shoot borer to continue as internode borer.

Rainfall during tillering and early part of cane formation and development stage (1-12 FAP) was negatively correlated with the incidence of scale insect (Table 2) as high rainfall events during crawler stage (nymphs) might have washed away the pest. Rainfall of above 500 mm limited the incidence of scale to approximately 10% and less than 400 mm rainfall during 11-12 FAP led to high incidence (Figure 4). Dry conditions might have predisposed the crop to scale insect activity as also observed by Gupta et al., [24]. The pest probably might have built up after the cessation of rains in the 18 FAP periods which corroborates the findings of Agarwal and Butani [25]. Rather than total rainfall, its pattern of distribution in a season appeared to be crucial in influencing scale insect dynamics. Fairly wide spread but non-beating rains coupled with high relative humidity from September onwards was found to influence the buildup of scale insect population [17].

Thermal Time

The data on the per cent incidence of ESB were regressed on accumulated thermal time in order to develop a predictive equation for ESB incidence.

Accumulated growing degree days was derived by using the formula [26].

$$\sum_{i=0}^{n}\left(\frac{Tmax + Tmin}{2}\right) - Tb$$

where,

T_{max} = maximum temperature (°C)
T_{min} = minimum temperature (°C)
T_b= base temperature (°C)

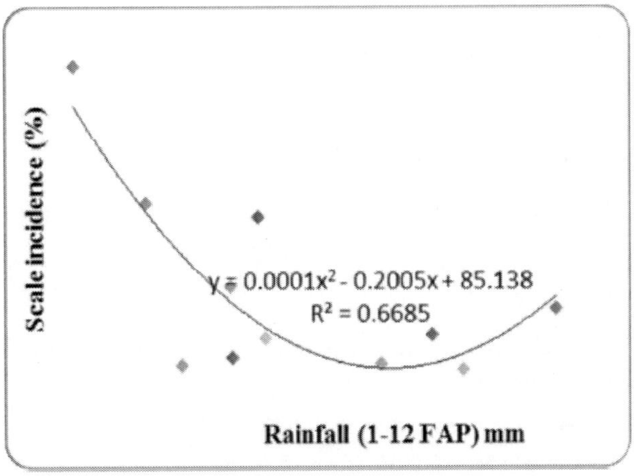

Figure 4. Relation between rainfall (mm) during 1-12 FAP period and scale insect incidence.

Table 2. Pearson's correlation coefficients between weather parameters and per cent scale insect incidence

FAP	Temperature (°C)				Relative humidity (%)			SSH	Rainfall (mm)	Evaporation (mm)	AGDD
	Max T	Min T	Mean T	Diurnal range	RH 1	RH 2	RH mean				
1	-0.55	-0.33	-0.48	-0.06	0.41	-0.50	-0.27	0.03	-0.40	-0.51	-0.48
2	-0.59	-0.40	-0.52	0.12	0.41	-0.35	-0.08	0.16	-0.07	-0.59	-0.52
3	-0.39	-0.13	-0.27	-0.20	0.618*	-0.43	-0.01	0.21	-0.19	-0.28	-0.27
4	0.00	-0.14	-0.09	0.15	0.40	-0.59	-0.48	0.27	-0.36	0.11	-0.09
5	-0.12	0.02	-0.06	-0.11	0.43	-0.19	-0.01	0.47	0.01	-0.03	-0.06
6	0.15	0.33	0.26	-0.15	-0.21	-0.39	-0.33	0.57	-0.11	0.38	0.26
7	0.71*	0.65*	0.73*	0.34	-0.45	-0.67*	-0.60	0.81**	-0.38	0.71*	0.73*
8	0.59	0.69*	0.73*	0.01	-0.83**	-0.61*	-0.74**	0.18	-0.30	0.67*	0.73*
9	0.41	0.53	0.56	-0.08	-0.21	-0.11	-0.15	0.44	0.18	0.46	0.56
10	0.14	0.56	0.65*	-0.29	-0.41	-0.21	-0.29	0.04	-0.25	0.23	0.65*
11	0.72*	0.64*	0.74**	0.01	-0.87**	-0.75**	-0.83**	0.39	-0.50	0.87**	0.74**
12	0.68*	0.59	0.74**	-0.18	-0.69*	-0.53	-0.65*	0.05	-0.36	0.83**	0.741**
13	0.19	0.45	0.39	-0.44	-0.10	0.17	0.09	-0.11	0.28	0.31	0.39
14	0.02	0.58	0.51	-0.59	-0.38	0.06	-0.03	-0.15	0.13	0.33	0.51
15	0.07	0.55	0.46	-0.51	-0.03	0.23	0.18	-0.13	-0.02	0.00	0.46
16	0.39	0.51	0.56	-0.35	-0.26	0.13	0.07	-0.09	0.30	0.05	0.56
17	0.79**	0.76**	0.85**	-0.45	-0.19	0.11	0.06	-0.01	-0.21	0.24	0.85**
18	0.18	0.77**	0.75**	-0.67*	-0.36	0.35	0.18	-0.47	0.79**	-0.28	0.75**
19	-0.03	0.67*	0.42	-0.62*	-0.66*	-0.13	-0.39	-0.40	-0.06	0.09	0.42
20	-0.08	0.55	0.26	-0.42	-0.72*	-0.43	-0.66*	-0.16	-0.18	0.46	0.28

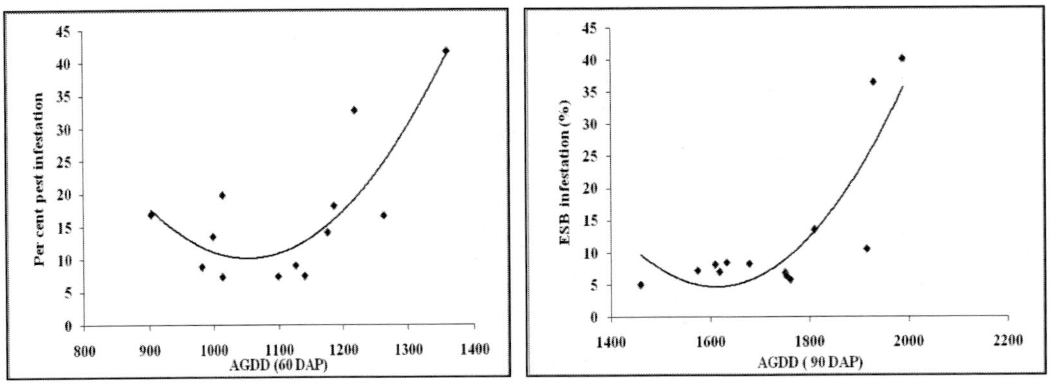

Figure 5. Incidence of ESB in relation to thermal time upto 60 and 90 DAP.

A base temperature of 10°C was considered for ESB and scale insect in the present analysis.

This analysis showed that ESB incidence accumulated around 1000 GDD by 60 DAP and 1600 GDD by 90 DAP (Figure 5). Considering the horizontal spread of pest incidence data, the authors restrained from proposing any functional relation to predict the incidence of the ESB as is done in case of several other pests like cashew leaf eating caterpillar [27], cotton bollworm [28]. The accumulated heat units during this period influenced the incidence of scale insect, *M. glomerata* (Figure 6).

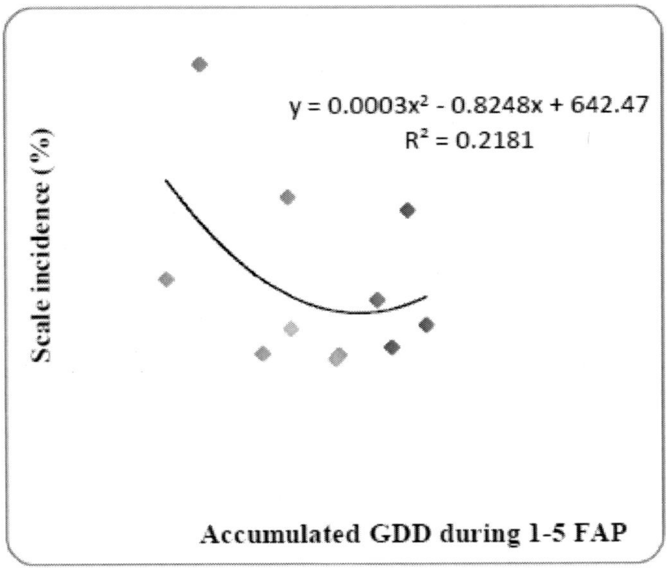

a

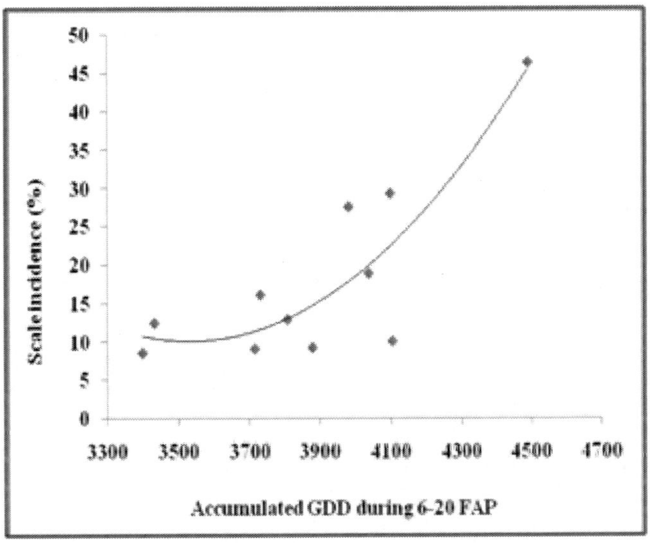

b

Figure 6. Relation between acumulated growing degree (AGDD)days and per cent scale insect incidence during 1-5 FAP (b) during 6-20 FAP.

This further strengthens the role of temperature on the population build up of the scale insect on sugarcane. The relationship between AGDD and scale insect was not strong in the early crop growth period (1-5 FAP) but the association strengthened during 6-20 FAP. At around 4500 AGDD highest scale incidence was noticed. The hours of bright sunshine (SSH) did not show any consistent effect but open pan evaporation showed positive influence during the grand growth stage of the cane. This can be expected as the open pan evaporation has a direct relationship with air temperature and as the temperature increases the evaporative demand of the air increases and thereby the evaporation (Table 2). Based on these correlation studies, regression analysis (linear and quadratic) was carried out with rainfall during 1-12 FAP and mean temperature during 7-12 FAP as independent variables and per cent incidence of *M. glomerata* as dependent variable.

The simple linear regression between mean temperature and *M. glomerata* resulted in a coefficient of determination (R^2) value of 0.64 (Table 2) thereby showing a fairly good account of variability of pest incidence due to mean temperature.

The regression of *M. glomerata* incidence on rainfall accounted for low influence of the weather factor ($R^2 = 0.35$). From Figure 6, it can also be deduced that the association between mean temperature and *M. glomerata* incidence is not linear necessitating a quadratic function fit through which R^2 value improved to 0.85 (Table 3).

Negative correlation of rainfall with scale infestation indicates that with increase in rainfall, the scale insect population decreases (Table 2). A quadratic fit with rainfall resulted in R^2 value of 0.67 (Table 3). When rainfall and mean temperature were together regressed against per cent incidence of scale insect, the coefficient of determination (R^2) was 0.64 which was lower than the individual effects. The mean relative humidity during 7 to 12 FAP accounted for 60 per cent of variation in the incidence of *M. glomerata*. The variable open pan evaporation was not included in the regression analysis because of its inconsistent effect throughout the crop growth period of sugarcane.

Accumulated growing degree days for initial 5 FAP and 6-20 FAP were regressed separately with per cent scale incidence and the equations are presented in Table 3.

The temperatures were low up to 5 FAP which is an indicative of fewer growing degree days and is unfavourable for scale insect. But later on, with increase in temperature scale incidence increased. The step-wise regression analysis resulted in the elimination of all weather variables except mean temperature of the 7-12 FAP and gave R2 value of 0.85 (Table 3). Based on regression analysis it can be concluded that temperature plays a major role in the incidence of scale insect on sugarcane in the north-coastal region of Andhra Pradesh, India.

The research on the relationship of weather parameters with the incidence of early shoot borer (*C. infuscatellus*) and scale insect (*M. glomerata*) on sugarcane indicated that early shoot borer incidence started from 40 DAP and the development and oviposition of ESB seems to be favoured by warmer (minimum temperature > 23.8°C) and dry nights (RH < 77 %). High rainfall events with rain exceeding 50 mm/day did not affect the spread of the ESB during the early stages of crop growth.

Table 3. Regression equations for scale insect based on weather parameters

Parameter	Regression equation	Coefficient of determination (R^2)
Mean temperature (7-12 FAP)	$Y=1148.28 - 84.34\ Tmean + 1.56\ Tmean^2$	0.85
AGDD during 6-20 FAP	$Y=799.4 - 0.44\ AGDD + 0.00006\ AGDD^2$	0.82
Mean relative humidity (7-12 FAP)	$Y=1257.06 - 30.697RH\ mean + 0.189RH\ mean^2$	0.75
Rainfall (1-12FAP)	$Y=85.22 - 0.201\ RF + 0.0001315\ RF^2$	0.67
AGDD during 1-5 FAP	$Y=684.6 - 0.88\ AGDD + 0.0002\ AGDD^2$	0.21

Prediction of ESB incidence using thermal time seems to be a remote possibility. It is evident from the present results that when high maximum temperature and low relative humidity prevails, *C. infuscatellus* is active as shoot borer during May-June and drought conditions coupled with low rainfall enable early shoot borer to continue as internode borer in North Coastal Region of Andhra Pradesh under rainfed conditions.

The activity of borer is reduced with the progress of monsoon rains. Light showers and cloudy weather are detrimental for the multiplication of C. *infuscatellus*.

The mean temperature and relative humidity during 7-12 FAP periods showed a significant association with the incidence of scale insect. As the mean temperature during 7-12 FAP increased, the scale insect incidence also increased. The relationship between AGDD and scale insect was not strong in the early crop growth period (1-5 FAP) but the association strengthened during 6-20 FAP. At around 4500 AGDD highest scale incidence was noticed. The hours of bright sunshine (SSH) did not show any consistent effect but open pan evaporation showed positive influence during the cane formation and development stage of the sugarcane. This can be expected as the open pan evaporation has a direct relationship with air temperature and as the temperature increases the evaporative demand of the air increases and thereby the evaporation. The rainfall during formation of internodes reduce the population buildup of scale insect on sugarcane. Prophylactic measures to control this pest need to be taken during 7-12 FAP period so that yield losses can be minimised.

The present research evidenced that among all the weather factors, temperature and relative humidity exerted more influence on the incidence of early shoot borer and scale insect in the North-Coastal Region of Andhra Pradesh and the relation developed from the present study can be used in the development of a decision support system to manage the early shoot borer and scale insect in sugarcane.

ACKNOWLEDGMENTS

The authors are thankful to the Director of Research, ANGRAU, the Associate Director of Research and the Principal Scientist (Sugarcane), Regional Agricultural Research Station, Anakapalle, Andhra Pradesh, India for providing facilities during the research work.

REFERENCES

[1] Krishnamuthty Rao, B. H. 1966. The activity of *Chilotraea infuscatellus* Snell. in relation to meteorological conditions. *Proceedings of Annual Convention of Sugarcane Technologists Association of India*, 34:127-129.

[2] Prasada Rao, V. L. V., Samdasiva Rao, S. and Venugopala Rao, N. 1991. Factors influencing infestation of early shoot borer (*Chilo infuscatellus* Snellen.) in Sugarcane. *Cooperative Sugar,* 22 (8): 515-521.

[3] Kalra, A. N. and N. C. Sharma 1963. Occurrence of the shoot borer, *Chilotraea infuscateus* Snell. as cane borer in Sriganganagar area of Rajasthan. *Indian Journal of Sugarcane Research and Development* 7: 193-194.

[4] Kalra, A. N. 1964. Recent advances in studies on influence of temperature and humidity on incidence and population of some of the major sugarcane pests. *Ent. Res. Wkrs. Conf.,* Delhi (Unpub.).

[5] Kalra, A. N. 1966. Sugarcane pests problem in India. *Proceedings of All India Symposium on Sugarcane Development,* Delhi, pp. 1-11.

[6] Rao, V. P. 1964. The sugarcane borers and other pest problems of old and their position in the present day plant protection in coastal Andhra. *Proceedings of All India Conference of Sugarcane Research and Development Workers*, 5: 547-550.

[7] Avasthy, P. N., Krishnamurthy, T. N. and Ananthanarayana, K. 1969. Factors affecting shoot borer in sugarcane. *World Crops*, 21: 39-40.

[8] Kalyanaraman, V. M., A. Leela David and P. S. Narayana swamy 1963. Distribution, seasons of occurrence and control of the early shoot borer of sugarcane, *Chilotraea infuscatellus* Snellen in Madras state. *Indian Journal of Sugarcane Research and Development* 7:89-95.

[9] DACNET, 2012. http://www.dsd.dacnet.nic.in/PestManage.htm.

[10] Sithanantham, Durai, S and Muthusamy, S. 1975. Incidence of sugarcane shoot borer in relation to planting time. *Indian Sugar*, 27:575-578.

[11] Avasthy, P. N. and Tiwari, N. K. 1986. The shoot borer *Chilo infuscatellus* Snellen. pp. 69-82. In: *Sugarcane Entomology in India* (eds.: David H. S. Eswaramoorthy and R. Jayanthi) Sugarcane Breeding Institute, Coimbatore, India.

[12] Rao, D. V. S. 1962. Studies on the resistance of sugarcane to the early shoot borer, *Chilotrea infuscatellus* Snell. *M.Sc. Thesis,* Andhra Univ. Waltair, pp. 132.

[13] Siva Rao, A. V. and Kamalakar Rao, C. 1963. Preliminary studies on some aspects of influence of certain climatic factors on borer population (*Chilotraea infuscatellus* Snellen) in sugarcane. *Indian Journal of Sugarcane Research and Development,* 7: 164-167.

[14] Mali, B. B. 1990. Studies on the seasonal incidence of early shoot borer Chilo infuscatellus Snell. in Vidarbha region. *Papers of the Fortieth Annual Convention of the Deccan Sugar Technologists Association* 1: 261-264.

[15] Rao, N. V. and Babu, T. R. 2004. Monitoring of the sugarcane early shoot borer, *Chilo infuscatellus* Snellen population by using light traps. *Journal of Entomological Research*, 28(3): 233-239.

[16] Shukla and Tripathi, 1983. Life history and seasonal history of sugarcane scale insect, *Melanaspis glomerata* (Green) (Hemiptera: Coccidae) in eastern U.P. *Indian J. Agric. Sci.,* 53: 160-162.

[17] Krishnamurthy Rao, B. H. 1977. Sugarcane scales and their control. *Sugarcane News*, 9: 57-71.

[18] Kalra, A. N. 1967. Studies on incidence and behavior of some major sugarcane pests in relation to weather and climatic conditions. *Indian Sugar* 17: 175-179.

[19] Tanwar, R. K. and Bajpai, P. K. 1993. Relative abundance of borer species in relation to environmental factors at shoot stage of sugarcane. *Indian Sugar,* XLIII: 243-248.

[20] Nagaraja Rao, P. R. and Chandy, K. C. 1957. Studies on the incidence of sugarcane borers. *Indian Journal of Sugarcane Research and Development* 2:23-30.

[21] Varadharajan, G. K., Saivaraj, K., Sathimoorthy, A. S., Subramaniam, A., and Kuppuswami, N. T. 1972. Sugarcane borers at Cuddalore (Tamil Nadu). *Indian Sug*., 21: 817-820.

[22] Hapase, D. G., Patil, A. S. and Moholkar, P. R. 1979. Effects of some climatic factors on the incidence of sugarcane borers. *Indian Journal of Sugarcane Technology*, 2: 1-8.

[23] Pradhan, S. and Bhatia, S. K. 1956. The effect of temperature and humidity on the development of the sugarcane stem borer, *Chilotraea infuscatellus* Snell. *Proceedings of International Society of Sugarcane Technologists*, 9: 856-869.

[24] Gupta, K. M., Pandey, B. N., Singh, R. A., Hans Nath and Dayal, R. 1976. Further spread of sugarcane scale insect (*Melanaspis glomerata* G.) in Uttar Pradesh. *Indian Sugar,* 26: 505-6.

[25] Agarwal, R. A. and Butani 1973. Sugarcane scale insect and its control. *Cane Grow. Bull.*, 1: 7-9.

[26] Iwata, F. 1984. Heat Unit Concept of crop maturity. In: *Physiological Aspects of Dry Land Farming*. Gupta, U. S. (Eds.) Oxford and IBH Publishers, New Delhi: 351-370.

[27] Sahu, K. R., Katlam, B. P. and Chaudhary, J. L. 2010. Impact of climatic factors on infestation of leaf eating caterpillar (*Mentrysia hyrtica*) of cashew in Chhattisgarh. *Journal of Agrometeorology,* 12 (1): 105-107.

[28] Mukherjee, A. and Bhoumik, P. 2009. Incidence of cotton bollworm (Helicoverpa armigera Hibner) in relation to meteorological parameters in the saline zone of West Bengal. *Journal of Agrometeorology,* 11 (2): 169-171.

In: Current Status of Sugarcane Research in India
Editors: A. K. Tiwari, M. Lal and A. K. Singh
ISBN: 978-1-63463-458-8

Chapter 2

STATUS OF SUGARCANE SCENARIO AND VARIETAL IMPROVEMENT PROGRAMME IN ANDHRA PRADESH

M. Charumathi [*] ***and K. Prasada Rao***
Division of Plant Breeding, RARS, Anakapalle, AP, India

ABSTRACT

Sugarcane is cultivated in both tropical and subtropical regions of India. In Andhra Pradesh about 4.0 lakh farmers were engaged in sugarcane production and 0.20 lakhs workers get direct employment in its processing for sugar. Maximum cane area is in coastal region followed by Rayalaseema and Telangana. In recent years, area under sugarcane has drastically reduced due to high production cost, scarcity of labour and stiff competition from other crops like maize, sunflower, soybean, ground nut and paddy. Cane area (2.64 to 1.96 lakhs ha), cane production (216.92 to 156.80 MT), cane crushed (173.23 to 103.00 lakh MT), sugar produced (16.80 to 9.93 MT) and cane productivity (82.20 to 80.00 MT) decreased from 2006-07 to 2012-2013. The cane yields have to be increased from present level of 80.00 t/ha to 90 t/ha by 2015 and 100 t/ha by 2020. Scientists should concentrate on development of clones with high yield potential, rich in quality and tolerance to biotic and abiotic stresses to step up cane yields by adopting medium and long term approaches.

Keywords: Sugarcane scenario in Andhra Pradesh -constraints –objectives –strategies to increase cane productivity

INTRODUCTION

Sugarcane, an annual crop, is cultivated in both the tropics and sub-tropics of India. It occupies about 3.0 per cent of the total cultivated area contributing about 7.5 per cent of the

[*] Corresponding author: Email: mcm_rars@yahoo.co.in.

gross value of agricultural production in the country. A major portion of sugarcane cultivation in India lies in the sub tropical belt but favorable agro-climatic conditions for its growth are available in the tropical belt. Because of that, the yields are substantially higher in the tropical belt as compared to the sub tropical regions. About 60 per cent of total cane production is utilized for sugar production, 30 per cent is consumed for producing *gur* and *khandsari* and the remaining 10 per cent is used for seed purposes. India is the world's largest producer accounting for nearly 15 per cent of the world's sugar.

Sugarcane is one of the important cash crops of Andhra Pradesh. It is cultivated in about 1.96 lakh hectares with a cane production 156.80 MT under assured irrigated, limited irrigated, rainfed conditions. Andhra Pradesh occupies 6th position in the country for cane area and production. However, cane productivity is low and stagnant in the state compared to neighboring states (Tamil Nadu, Karnataka and Maharastra). Increase in cane area in marginal soils, rainfed conditions, moisture stress during formative phase, non adoption of recommended package of practices in plant and ratoon crops are some of the major constraints of cane production.

The net returns obtained from cane cultivation are low and marginal due to increase in cost of production particularly harvesting charges and low cane yields obtained from the unit area. The cane area in the state declined from 2.64 lakh hectares (2006-07) to 1.96 lakh hectares (2012-13). The cane area is being diverted to more profitable alternate crops viz., maize, sunflower, cotton, soybean, ground nut and paddy in many traditional cane growing districts.

Sugarcane Scenario at State Level

Being an important commercial crop of Andhra Pradesh, sugarcane provides raw material to the sugar industry which is an important agro -based industry in the state. It is substantially contributing to rural economic growth. About 4.0 lakh farmers are engaged in sugarcane production and 0.20 lakhs workers get direct employment in its processing for sugar. About 1.5 lakh people derive their livelihood from small scale cottage jaggery industry. Besides, the sugarcane crop is a host of other gainful employment in other industries which use its by-products as the raw material for alcohol and alcohol based industries. Sugar industry pays Rs 1500 crores to sugarcane growers every year and Rs 40 crores by way of excise duty and about Rs 50 crores as purchase tax and cess on cane.

Sugarcane is cultivated in a wide range of situations *viz.,* assured irrigated, limited irrigated and rainfed situations. Andhra Pradesh ranks 6th position in cane area and production in the country. It was grown in about 1.96 lakh ha with a production of 156.80 lakh metric tons. The average cane yield was 80.00 t / ha (2012-13).

There was a progressive increase in cane area from 0.91 (1960) to 1.96 lakh hectares (2012-13) in the state (Table 1). Three fold increase in cane production from 67.48 (1960-61) to 156.80 lakh MTs (2012-13) was recorded during the same period. However, the cane productivity has been stagnant (72 – 82 t/ha) for the last four decades. Therefore, new approaches need be adopted for increasing the cane productivity.

Table 1. Cane area, production, productivity in Andhra Pradesh

Year	Cane area (Lakh ha)	Cane production (Lakh MT)	Cane yield (t/ha)
1960-61	0.91	67.48	74.00
1970-71	1.20	91.22	76.00
1980-81	1.32	101.86	77.20
1990-91	1.82	126.69	69.60
2000-01	2.17	176.90	81.40
2001-02	2.18	180.82	82.90
2002-03	2.32	153.87	66.20
2003-04	2.09	150.70	72.10
2004-05	2.10	157.39	74.90
2005-06	2.30	176.56	76.80
2006-07	2.64	216.92	82.20
2007-08	2.47	202.04	82.00
2008-09	1.96	153.22	78.00
2009-10	1.58	117.08	74.10
2010-11	1.92	147.84	77.00
2011-12	2.03	142.10	70.00
2012-13	1.96	156.80	80.00

Quantity of cane crushed in sugar factories varied from 19.33 (1960-61) to 103.00 lakh MT (2012-13). Similarly, sugar production ranged from 1.83 (1960-61) to 9.93 lakh MT (2012-13). The number of sugar factories in the state increased from 12 to 36 in the same period with a total installed crushing capacity 1.00 lakh MT/day. Maximum numbers of sugar factories are in private sectors. The crushing capacity ranges from 1000 to 8500 tonnes crushing per day.

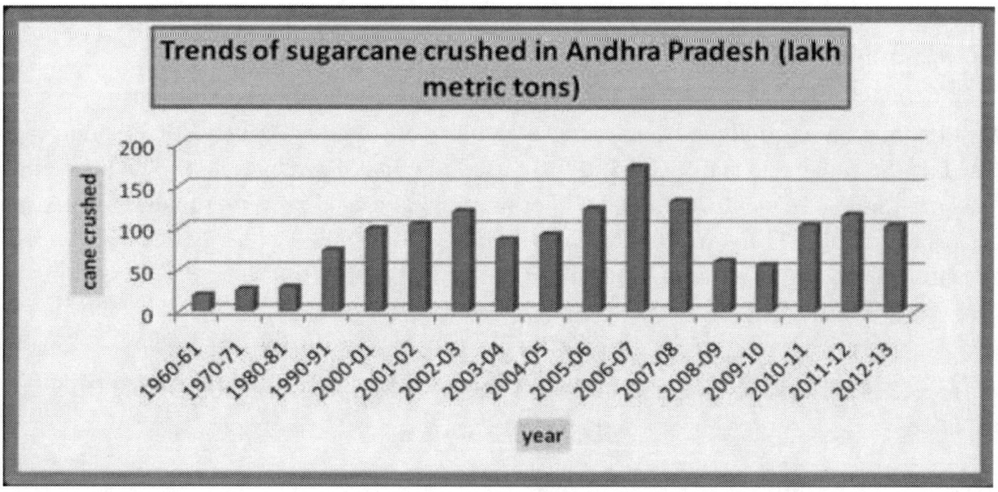

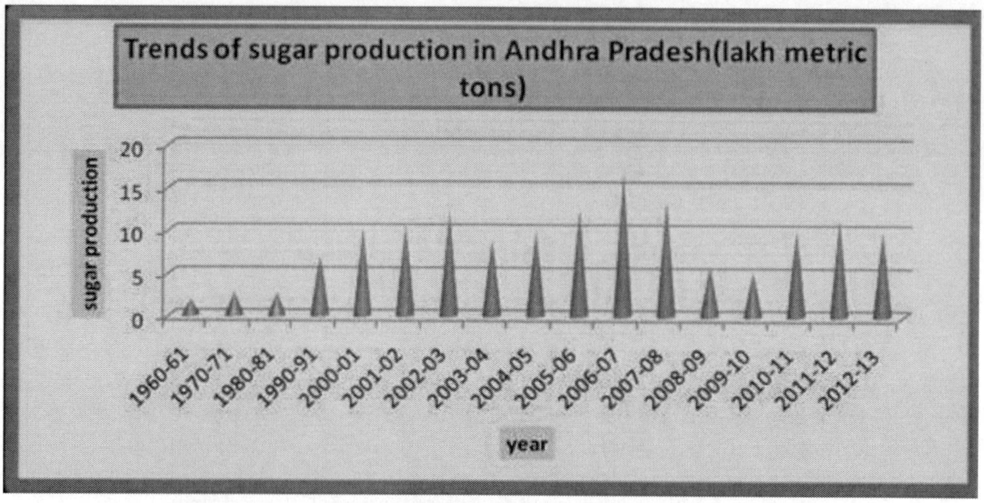

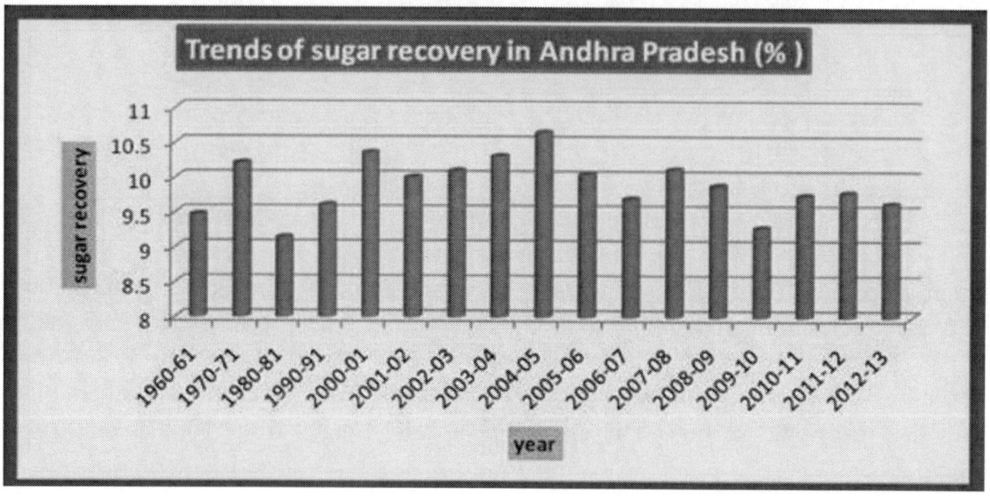

Fifteen sugar factories have cogeneration units attached to sugar factories with an installed capacity of 171.15 MW in Andhra Pradesh. Five sugar factories have ethanol units with an installed capacity of 215 KLPD. Molasses is the product of sugar industry. About of 40 kg of molasses is produced per tonne of cane from which 10 liters of ethanol can be obtained. If the sugarcane is directly and fully used in ethanol production, the yield of ethanol is 70 ltr per tonne of cane. Ethanol is an alternate fuel and is a basic raw material for various chemicals and liquor. There are about 25 distilleries with installed capacity of 1,53,282 KLPD (1,900 million ltr/year) in Andhra Pradesh.

Causes for Reduction in Cane Area and Production in the State

1. Increase in cost of production of cane
2. Scarcity of labour coupled with high labour wages
3. Low and non – remunerative price paid by the sugar factories

4. Stiff competition from other alternative crops like maize, sunflower, soybean, ground nut and paddy
5. Non availability of labour
6. Stagnation in cane productivity
7. Delayed payment of cane price

MANDATE

- Evolving high yielding and sucrose rich clones possessing tolerance / resistance to biotic and abiotic stresses, good ratoonability and suitability for different agro-climatic zones.

CONSTRAINTS OF CANE PRODUCTION

1. Moisture stress during formative phase coinciding with summer
2. Water logged conditions during September-October months
3. Crop lodging due to cyclones during North East Monsoon
4. Extension of cane area to marginal and sub marginal lands
5. Increase in cane area under rainfed condition
6. Incidence of insect pests and diseases
7. Poor ratoon cane management

PRESENT LINE OF WORK OF SUGARCANE IN DIFFERENT RESEARCH STATION IN ANDHRA PRADESH

- Identification of sugar rich clones of short and mid duration
- Clones tolerant to moisture stress and quality deterioration
- Clones tolerant to biotic stresses like red rot, smut, and YLD
- Clones tolerant to ESB, Internodal borer, Scale and other sucking pests
- Rain fed sugarcane production technologies
- Low cost sugarcane production technologies like bud chip planting, spaced planting, irrigation methods which include surface and sub surface drip along with fertigation
- Standardization of micropropagation protocol for mass production of sugarcane planting material
- Sugarcane inter cropping
- Organic farming in sugarcane with different organics and factory effluents
- Verification of STCR equation for fertilizer recommendation
- Evaluation of varieties for post harvest deterioration and jaggery
- Modules for powder/ granular jaggery
- Jaggery research using organic farming, herbal clarificants and fortification with vitamins and minerals

Table 2. Progress of fluff supply programme from 1998-99 to 2012-13 at RARS; Anakapalle

Year	Quantity of fluff received (g)	No. of crosses/GCs/ PCs studied			No. of seedlings			No. of genotypes selected/ evaluated in		
		Crosses	GCs	PCs	Transplanted	Survived	per cent Survival	Seedling nursery (C_0)	Settling nursery (C_1)	Selection nursery (C_2)
1998-99	3,509.42	41	4	18	38095	23424	61.49	210	50/168	12/20
1999-00	5,353.40	54	17	19	32153	25904	80.56	318	58/210	12/50
2000-01	1,136.98	23	12	-	3332	2735	82.08	103	86/318	16/58
2001-02	2,804.70	47	25	7	13711	10226	74.58	252	20/101	20/86
2002-03	2,719.50	34	24	8	22303	11245	50.42	315	38/252	11/20
2003-04	1,329.00	23	21	11	11869	7590	63.95	131	62/315	16/38
2004-05	1,698.90	24	42	5	12389	9792	79.04	175	30/131	23/62
2005-06	1136.65	29	41	-	31235	12152	38.91	317	44/175	11/23
2006-07	1177.99	39	29	-	15424	11560	74.95	220	40/317	24/44
2007-08	1313.40	46	19	-	17311	13692	79.61	520	52/220	17/40
2008-09	1744.39	49	37	10	15005	9193	61.27	472	114/520	18/52
2009-10	1102.70	42	19	4	9588	5260	54.86	519	66/472	23/114
2010-11	1748.77	40	35	12	14337	4537	31.65	321	100/519	21/66
2011-12	1941.22	54	38	12	16228	11620	71.60	357	91/321	41/100
2012-13	1142.99	30	23	13	14213	6250	43.97	369	104/357	25/91
Total	29860.01	575	386	119	267193	165180	63.26	4599	955	290

Table 3(a). Promising clones under Zonal Varietal Trials (Early) during 1998-99 to 2012-13

S. No.	Clone No.	Parentage	Cane Yield (t/ha)				CCS Yield (t/ha)				Juice Sucrose Percent				Reaction to diseases	
			I Plant	II Plant	Ratoon	Mean	I Plant	II Plant	Ratoon	Mean	I Plant	II Plant	Ratoon	Mean	Red Rot	Smut
1.	CoA 95081 (90A272)	CoA 7602xCoT8201	114.32	119.36	85.42	106.37	13.40	13.81	10.52	12.58	16.68	16.03	16.86	16.52	R	R
2.	CoA 98081 (92A60)		113.54	108.71	86.19	102.81	12.91	13.06	11.65	12.54	16.13	18.46	18.67	17.75	R	MS
3.	CoA 98082 (92A355)	CoT8201GC	128.50	103.04	71.60	101.05	14.65	13.33	10.92	12.97	16.04	17.75	15.69	16.49	R	HS
4.	CoA 99081(93A53)	Co6304x CoT8201	113.19	98.84	63.37	91.80	13.66	10.24	8.67	10.86	18.96	18.59	18.64	18.73	R	S
5.	CoA 99082 (93A145)	CoT 8201 x B38192	128.47	112.27	90.52	110.42	13.46	13.79	13.30	13.52	18.89	16.48	20.20	18.52	R	HS
6.	CoA 01081(96 A 176)	Co 86062 GC	98.01	115.70	80.32	98.01	12.80	15.26	10.31	12.79	19.10	18.93	19.24	19.09	R	MS
7.	CoA 02081(94 A 124)	MS 6847 GC	97.01	115.00	90.00	100.67	10.87	13.52	12.08	12.16	15.27	16.94	17.23	16.48	R	MS
8.	CoA 03081(97 A 85)	Co 8212 GC	112.50	115.00	98.08	108.53	14.28	14.41	12.46	13.72	17.51	17.50	17.61	17.54	R	MS
9.	CoA 05321(2000 A213)	Co 740 PC	129.67	132.60	100.30	120.86	17.95	16.61	12.70	15.75	18.63	18.15	18.23	18.34	R	MS
10.	CoA 06321(2001 A 63)	86 A 146 GC	126.30	136.00	98.00	126.30	17.79	17.02	12.57	17.79	17.98	17.51	17.88	17.98	R	MS
11.	CoA 07321 (2000A56)	87A298xHR83-65	133.67	138.50	97.25	123.14	15.93	16.69	11.67	14.76	16.68	17.34	17.41	17.14	R	R
12.	CoA 08323(2003A255)	80R41GC	112.00	138.00	103.67	117.89	13.26	17.37	13.10	14.58	16.83	17.80	18.23	17.62	R	MS
13.	CoA 09321 (2004A55)	Co 86002 x Co 92008	136.67	125.33	98.00	120.00	16.32	16.05	12.94	15.10	17.20	18.00	18.30	17.83	R	MS

Table 3(b). Promising clones in Zonal Varietal Trials (Mid late) during 1998-99 to 2012-13

S. No.	Clone No.	Parentage	Cane Yield (t/ha)				CCS Yield (t/ha)				Juice Sucrose Percent				Reaction to diseases	
			I Plant	II Plant	Ratoon	Mean	I Plant	II Plant	Ratoon	Mean	I Plant	II Plant	Ratoon	Mean	Red Rot	Smut
1.	CoA 90081 (87A380)	CoC 671xCoA 7602	112.02	113.05	74.43	99.83	15.00	14.44	10.23	13.22	18.00	17.70	18.93	18.21	R	S
2.	CoA 94081 (87A397)	Co7201xCo775	125.00	128.13	77.34	110.39	15.56	16.56	10.24	14.12	17.80	17.99	18.02	17.94	R	S
3.	CoA 94082 (88A90)		117.02	117.00	69.93	101.32	14.50	14.80	8.32	12.54	17.30	17.42	16.78	17.67	R	HS
4.	CoA 93082 (88A162)	Co78201xCP 44-101	125.67	127.78	76.37	109.94	15.80	16.38	9.92	14.03	17.40	17.89	18.01	17.77	R	S
5.	CoA 02082 (96 A 136)	Co 97027 GC	111.72	115.54	78.87	102.04	15.36	15.57	10.83	13.92	18.99	19.01	19.10	19.03	R	HS
6.	CoA 97081 (90 A 278)	Co A 7602 x CoT 8201	98.06	100.00	76.34	97.81	12.00	11.94	9.07	11.00	17.01	16.53	17.06	16.87	R	MS
7.	CoA 2000 - 081 (93 A 11)	Co 8318 GC	110.20	108.15	73.89	97.41	16.00	15.19	10.39	13.86	19.00	19.28	18.68	18.99	R	MS
8.	CoA 2000-82 (94 A 109)	Co A 7602 GC	99.50	89.63	58.43	82.52	11.53	11.54	7.31	10.13	17.49	17.18	17.63	17.43	R	MS
9.	CoA 01-082 (96 A 3)	85 A 261 x Co A 7602	117.50	109.17	80.14	102.27	17.56	15.87	11.84	15.09	20.59	19.68	21.62	20.63	R	MS
10.	CoA 03082 (97 A44)	MS 6847 x Co 775	93.93	95.00	55.20	81.38	11.98	12.10	11.13	11.74	17.77	17.87	18.00	17.88	R	MS
11.	CoA 05322 (98 A 163)	Co 7706 x Co 6904	126.00	133.67	91.30	116.99	16.63	16.30	11.50	14.81	18.23	18.22	18.00	18.15	R	MS
12.	CoA 05323 (2000 A 225)	Co 85002 PC	128.00	123.00	91.30	114.10	16.65	15.40	11.50	14.52	18.00	18.95	19.00	18.65	R	MS
13.	CoA 6322 (2001 A 85)	Co A 7602 PC	122.00	123.00	91.30	112.10	16.41	15.40	9.70	13.84	17.73	18.95	18.84	18.51	R	MS
14.	Co A 07322 (2000A241)	79A28 x CoA 7602	100.00	107.00	89.25	98.75	12.86	12.91	10.73	12.17	17.99	18.40	18.11	18.17	R	MS

Varieties Developed at Rars, Anakapalle

S. No	Clone	Parentage	Year of Release	Maturity group	Cane Yield (t/ha)	Juice Sucrose (per cent)	Special features
1	Co 62175	Co951 x Co419	1968	Late	125 – 130	16.0	High yielder good jaggery variety, but susceptible to Red Rot
2	CoA 71-1	Co 1077 GC	1971	Early	105	18.0	Good jaggery variety, good tillering variety with high cane yield.
3	CoA 7601	Co 678 x Co 775	1976	Early	105	18.5	Short duration, high nitrogen use efficiency.
4	CoA 7602	Co1287 x Co775	1976	Midlate	100	16.0	Suitable for water logging, moisture stress conditions and resistant to red rot
5	CoA 7701	Co 419 x Co 62174	1978	Early	95	18.5	Good tillering ability, good retaining ability, Resistant to red rot
6	Co 7508	Co 62174 GC	1981	Early	95 – 100	19.20	Rich in juice sucrose, Resistant to red rot
7	CoA 8401	Co 6304 x Co 1287	1989	Early	90	17-18	Good tillering ability, good retaining ability, Resistant to red rot
8	Co 7706	Co 740 x Co 775	1989	Late	120.15	16-17	Good tillering ability, good retaining ability, Resistant to red rot .Good jaggery quality variety,
9	85A261	Co 6806 x Co 775	1996	Early	100-105	19-20	Rich in juice sucrose, Resistant to red rot
10	Madhu (84A125)	CoC 671 x Co 6304	2002	Early	100-110	18-19	Suitable for water logging and moisture stress conditions.
11	Viswamithra (87A298)	Co7704 x CoC 671	2002	Early	110-120	18-19.5	Suitable for all types of soils, Rainfed, water logged situations, good ratooning ability, resistant to red rot and susceptible to smut
12	Sarada (93A145)	CoT 8201 x B 38192	2006	Early	115-120	17.5-18.0	Suitable for water logged, rain fed limited irrigated and saline alkali soils. Resistant to red rot.
13	Visakha (97A85)	Co 8212 GC	2010	Early	120	17.5	Suitable for early planted, late planted rainfed conditions, moisture stress conditions and limited irrigated situations. Resistant to red rot and smut.
14	Kanakamahalakshmi (2001A63)	86A 146 GC	2012	Early	120-125	17.00	Suitable for early planted, late planted rainfed conditions, resistant to red rot.
15	Uttara (98A163)	Co 7706xCo6904	2012	Midlate	125-130	17.52	Good ratooning ability, resistant to red rot and moderately susceptible to smut

VARIETAL IMPROVEMENT PROGRAMME IN ANDHRA PRADESH

Varietal improvement in sugarcane is carried out through fluff supply programme and identification of location specific clones through zonal varietal trials. Profuse flowering and seed set takes place only in Coimbatore due to favourable agro-climatic conditions. Hybridization is carried out at the National Hybridization Garden, Sugarcane Breeding Institute, Coimbatore during October – December months. A large number of seedlings are raised every year from the fluff obtained from Sugarcane Breeding Institute, Coimbatore to select desirable genotypes. Multi stage selection is carried out through seedling, settling, preliminary yield trials, main yield trials (plant and ratoon crops) and on-farm trials to identify improved clones with high cane yield, sucrose, good ratoonability and tolerance to abiotic and biotic stresses. It takes about 8-10 years to evolve a clone through fluff supply programme. Twenty eight clones were developed/identified so far by all four sugarcane research stations of Acharya N.G. Ranga Agricultural University and released for commercial cultivation in the state. Clones developed at various sugarcane research stations in East Coast Zone and Peninsular Zone are also tested under zonal varietal trials at all the sugarcane research stations of Acharya N.G. Ranga Agricultural University to select superior clones.

A total of 2,67,193 seedlings from 575 crosses, 386 GCs and 119 PCs obtained from Sugarcane Breeding Institute, Coimbatore were raised from 1998-2012. Out of those, 955 selections were made in settling nursery and 290 selections were made in selection nursery (Table 2). Based on agro climatic conditions the country is divided into five zones viz., Peninsular zone, East coast zone, North West zone, North central zone and North east zone. East coast zone comprising Andhra Pradesh, Tamil Nadu and Orissa is the most varied zone that helps in identifying the best clones suited for the zone. The clones performing consistently across the locations in two plant crops and one ratoon crop in advanced varietal trials with in the zone are recommended for commercial cultivation. A total of 13 early and 14 midlate clones have been evolved from zonal varietal trials, in which clones 87A298, 90A272, 93A145, 97A85, 2000A56, 2001A63, 2003A255, 2004A55, 87A380, 98A163 and 2000A225 are popular in cane growing areas in the state (Table 3a and 3b). Out of these, 87A 298 is the most popular variety occupying an area of 65-70 percent in Andhra Pradesh, Tamil Nadu and Karnataka.

97A85

87A298

93A145

CoA 7602

Co 7706

Co 7219

Co 6907

81V48

98A163

2003A 255

2004A 55

2000A 225

2001A63

83R23

91V83

Thrust Areas

1. Developing of high yielding sucrose rich varieties for stabilizing cane yields and sugar production.
2. Evolving varieties suitable for co-generation (high bio mass and moderate juice sucrose) and ethanol production (maximum juice extraction and moderate juice sucrose).
3. Varieties with wide genetic base of resistance to diseases like red rot, smut and grassy shoot
4. Versatile varieties having a broad spectrum of inherent abilities suitable for situations like rainfed, water logging and problematic soils
5. Varieties with high clonal efficiencies for photosynthesis, high conversion rate of biomass to productive components
6. Varieties capable of withstanding moisture stress during maturity phase with prolonged shelf life for quality and yields
7. Utilization of tissue culture propagules for rapid exploitation of potential new varieties in extensive areas
8. Development of varieties capable of accumulating sugar irrespective of prevailing weather conditions especially before onset of winter
9. Technologies for improving productivity in rainfed sugarcane
10. Identification of clones tolerance to major diseases at the early stages of selection adopting biotechnology tools
11. Perfection of mechanization of granular jaggery production
12. Nutrient requirements of the varieties in relation to the soils and farming situations (Rainfed, ID, stagnation, ill drained, problematic conditions in relation to soil type)
13. Identification of clones with high nutrient utilization efficiency (maximum uptake) and response to lower doses of fertilization
14. Designing the most efficient farm implements / machinery for cane cultivation
15. Mass production of bio-control and botanical agents against major insect pests and diseases
16. Studies to spot out hormones/growth promoters that activate physiological processes in sugarcane ratoons
17. Molecular characterization of red rot isolates by PCR techniques
18. Development of clones suitable for mechanical harvesting.

Approaches for Increasing Cane Productivity

A. Short Term Approach

S. No	Constraint(s)	Critical intervention(s)
1.	Salinity	• Leaching out excess salts with good quality irrigation water. • If good quality irrigation water is not available, saline soils can be reclaimed making use of rain water during rainy season alone

S. No	Constraint(s)	Critical intervention(s)
2.	Alkalinity	• Application of gypsum depending upon soil pH • Incorporate gypsum in soil by disk plough after letting water. Allow to settle for 24 hours and drain soluble sodium salts. • Raising of daincha and incorporate into soil at 50 per cent floral bud production stage and then grow paddy for first on season, incorporate paddy straw and raise sugarcane. • Cane yield can be increased by 50 per cent in saline soils and 150 per cent in alkali soils by reclamation
3.	Surface crusting in chalka soils	• Add of paddy husk or powdered groundnut shells @ 2.5 t / ha and incorporate into soil • Providing irrigations at closer intervals and do not allow the soil to dry
4.	Hard pan formation in heavy soils	• Deep ploughing or chistelling or subsoiling will improve root growth, water uptake and anchorage to crop and reduces bulk density. • Deep ploughing helps to increase cane yields by 8 per cent
5.	Water logging in heavy soils	• Formation of drainage channels at 12-15 mts interval • Draining of excess water with the help of archemedian screw, swing basket / low lift pump • Cultivation of clones viz., Co 6907, 84 A 125, 86 V 96, Co 7219, Co 7805, Co T 8201, 83 V 15, 83 V 288, 87 A 298, Co A 7602, 91 V 83, Co 7706 • Early planting (January) • Heavy earthing up • Early application of fertilizer (at 30 and 60 DAP) • Foliar nutrition under ill drained conditions (urea + MOP @ 2.5 per cent) • Management of white fly
6.	Moisture stress during formative phase	• Growing drought tolerant varieties viz., Co 6907, 81 A 99, 84 A 125, 87 A 298, Co A 7602, 83 A 30, 83 R 23, Co T 8201, 97 A 85 • Trash mulching @ 3 t / ha on 3rd day after planting • Adoption of high seed rate • Application of organic manures (FYM / PMC / Vermi compost / Trash compost) • Soaking setts in 10 per cent lime solution for one hour • Foliar spray with urea and MOP each @ 2.5 per cent • Basal application of potassium • Adoption of drip irrigation • Alternate furrow irrigation
7.	Low germination	• Use of planting material from short crop (raised from hot water – 52°C treated mature crop planting material) and maintenance of purity of plant material enhances cane yield by 15 per cent.
8.	Low water holding capacity of light soils	• Application of 25 t / ha of FYM or 12.5 t / ha pressmud cake or raising green manure crops like Sesbania. *Sunhemp*, Pillipesara, cowpea, daincha etc., and incorporating into the soil at 50 per cent floral bud production stage.
9.	High irrigation water requirement	• Sugarcane has to be irrigated at weekly intervals during formative phase and at 18 days interval during maturity phase. • Summer irrigations boost cane yields by at least 10 per cent

(Continued)

S. No	Constraint(s)	Critical intervention(s)
10.	Lodging of crop	• Deep ploughing and shallow planting in deep furrows • Working with mould board plough either way twice on the beds to throw the soil towards the sugarcane clump to provide proper anchorage and also to loosen the soil for wide root spread, better uptake of nutrients and rapid growth during grand growth period. Improves cane yield by 8 per cent. • Trash twist propping is a better practice than stooking as the former facilitates free aeration besides better anchorage to withstand wind blow. Improves cane yields by 3 per cent.
11.	Improper cane harvest	• Cane has to be harvested after attaining maturity • Harvesting of cane close to the base with sharp knives. Enhances income to the farmers by 3 per cent.
12.	Incidence of striga	• Hand weeding before flowering and seed formation and their destruction • Trap cropping – growing fodder jowar and removing striga plant along with it before flowering and seed formation • Spraying of 1-2 kg of 2, 4 - mixed in 500 litres of water per hectare
13.	Protracted period of planting	• Adoption of staggered planting facilitates harvesting of physiologically mature crop, increases cane and sugar yields.
14.	Insect pests and diseases	• Adoption of integrated insect and disease management practices
15.	Incidence of weeds	• Weeds reduce cane yields ranging from 15-70 per cent. The crop has to be kept weed free upto 90 days after planting.
16.	Poor ratoon cane management	• Stubble shaving, gap filling, off bearing, additional dose of N, trash mulching and correction of micro nutrient deficiency
17.	Non adoption of location specific clones	• Selection of location specific / farming system specific clones
18.	Non replenishment of soil nutrients	• Addition of organic sources viz., FYM, PMC, Vermi compost, trash compost and incorporation of green manure crops and application of inorganic fertilizer based on soil test values
19.	Non adoption of sett treatment	• Sett treatment with carbendazim (0.5 g / lt) and malathion (2 ml /lt) for 15 minutes will help in arresting sett transmitted diseases.
20.	Non adoption of proper varietal schedule	• Adoption of proper varietal schedule (70 per cent early and 30 per cent mid late clones) helps in increasing cane yield as well as sugar recoveries
21.	Difficulty in the application of insecticides and pesticides on standing crop	• Eco friendly management of pests and diseases of sugarcane using antagonists, bio-pesticides and parasites.
22.	Introduction of diseases/insect pests from neighbouring states	• Strict following of internal quarantine regulation at state level. • Indiscriminate introduction of a variety without phyto-sanitary certification should be a compulsion. Such measures are essential to check the production of red rot from neighbouring states.

B. Medium Term Approach

S. No	Approaches	Time frame (years)
1.	Rapid production of healthy planting material through micro propagation	3 – 4
2.	Identification of suitable clones through Zonal Varietal Testing	3 – 4
3.	Exploring possibilities for crossing programme at Madanapalle / Araku areas	3 – 4
4.	Screening of sugarcane clones for quality jaggery	3 – 4
5.	Working out of optimum water requirement in relation to age of crop, atmospheric temperature, humidity and wind flow conditions	3 – 4
6.	Standardization of fertigation schedule for sugarcane	
7.	Nutritional requirement for sugarcane under wide row planting	3 – 4
8.	Revision of fertilizer schedule based on soil test results	3 – 4
9.	Evolving of ICM practices to reduce cost of production	3 – 4
10.	Research on exploitation of potential of micro nutrients	3 – 4
11.	Research on effective utilization of distillery effluents	3 – 4
12.	Establishment of trash decomposing culture unit	3 – 4
13.	Production and supply of bio-agents against scale, whitefly, mealy bug, while grub and red rot	3 – 5
14.	Enhancement of parasitoid efficiency though info-chemicals against sugarcane pests.	
15.	Testing and evaluation of farm implements	3 – 4
16.	Standardization of jaggery powder making unit	3 – 5
17.	Development of non harmful clarificants for making jaggery	3 – 4
18.	Design and development of dryers for granular jaggery	3 – 4
19.	Design and development of batch type dryers for granular jaggery	3 – 4
20.	Production of value added products viz., tetra packed cane juice, flavored jaggery and syrup from cane juice, liquid jaggery, cane juice wine and commercial production of vinegar from sugarcane to boost the demand for cane and improve the profitability of sugarcane	3 – 4
21.	Organic jaggery	3 – 4
22.	Establishment of agro processing centres for transfer of technologies on value addition and farm machinery	3 – 4

C. Long Term Approach

S. No	Approaches	Time frame (years)
1.	Development of high yielding, sucrose rich clones possessing tolerance to biotic and abiotic stresses and good ratoonability.	6 – 8
2.	Evolving of improved clones with high yield potential for late planted rainfed conditions	6 – 8
3	Reduction in the varietal development period by cutting short the duration of varietal development through molecular markers	5 – 6
4.	Identification of varieties for specific needs of the sugar industry viz., Co-generation and ethanol production	6 – 8

(Continued)

5.	Elimination / rectification of defects in the old clones for a single character through bio technological approaches.	6 – 8
6.	Evolution of short duration clones possessing thermo-insensitivity to prolong crushing period.	6 – 8
7.	Utilization of markers in the varietal development to cut short the duration varietal development	8 – 10
8.	Design, fabrication and development of farm implements suitable for use in cane cultivation	5 – 6
9.	Development of transgenics for biotic & abiotic stresses	6 – 8
10.	Characterization of variants of red rot pathogen	6 – 8
11.	Selection of Physiological & biochemical parameters in relation to growth, maturity & ripening.	6 – 8
12.	Basic studies on identification of races of rust and its management	4 – 5
13.	Basic studies on grassy shoot disease	5 – 6
14.	Studies on yellow leaf streak disease	5 – 6
15.	DNA based molecular markers for assessing genetic purity of clones / regenerants	5 – 6

Mobilization of Resources for Sugarcane Research

An amount of five rupees per ton of cane crushed may be collected from sugar factories for research cum extension projects. In addition Rs. 1 / q of jaggery traded may be collected as 30 per cent of cane produced in the state is converted into jaggery and the jaggery farmers also being served with improved varieties and technologies. The funds should be available to meet the expenditure on R & D projects and help in hastening the productivity and production of sugarcane in the state of Andhra Pradesh.

Impact Analysis

Varieties and technologies recommended from the research stations have been accepted and adopted by farmers whether for the making of gur or white sugar by industry. Scientists of sugarcane research stations have come to the rescue of farmers and industry by releasing high yielding, sucrose rich varieties resistant to red rot and other abiotic stresses. Sugarcane research stations have substantially helped the jaggery making farmers with improved technologies in various aspects of jaggery, an important rural industry.

CONCLUSION

In Andhra Pradesh, a major portion of sugarcane area is under limited irrigated and rainfed conditions. That is why cane productivity is low and stagnant, due to an increase in area in marginal soils, rainfed conditions, and moisture stress during formative phase. Non adoption of recommended package of practices in plant and ratoon crops are some of the major constraints of cane production. Presently, few varieties are under cultivation over large diverse regions across the zone. There is a need to adopt new improved location specific varieties by adopting short, medium and long term approaches for realizing improved cane productivity, increased sugar recovery, along with extended crushing periods to make the running sugar factories viable and economical. In addition, quality and production of recommended varieties and seed multiplication as per requirements are to be given priorities for realizing the desired cane production. Announcement of remunerative statutory minimum price by the Government of India and introduction of partial mechanization in cane cultivation would further help in achieving the improved cane yields.

In: Current Status of Sugarcane Research in India
Editors: A. K. Tiwari, M. Lal and A. K. Singh
ISBN: 978-1-63463-458-8

Chapter 3

EMBRACING BIOTECHNOLOGY METHODS FOR CROP IMPROVEMENT RESEARCH IN SUGARCANE

Rachayya M. Devarumath[1*], Gauri A. Nerkar[1], Forough J. Farsangi, Ashok A. Nikam[2] and K. Harinath Babu[1]
[1]Molecular Biology and Genetic Engineering Division,
[2]Tissue Culture Section, Vasantdada Sugar Institute, Manjari (Bk), Pune, India

ABSTRACT

Sugarcane (*Saccharum officinarum* L.) is one of the most important field crops grown in the tropics and sub-tropics. More than half of the world's sugar is derived from sugarcane. Conventional methods have greatly contributed to crop improvement; however limitations such as complex genome, narrow genetic base, poor fertility, susceptibility to biotic and abiotic stresses and long duration to breed elite cultivars still impose a challenge. Sugarcane, thus, is the suitable candidate for application of biotechnology and genetic engineering tools. In this direction, *in vitro* culture systems and related biotechnological tools have been developed as novel strategies for sugarcane improvement. Studies have been conducted towards employing *in vitro* culture combined with radiation/chemical induced mutagenesis for mutant isolation. Advancements in genomics tools have paved the way for a detailed understanding of the mechanism underlying biotic and abiotic stress responses. The potential of the current genomics programs, aimed at elucidating the structure, function, and interactions of the sugarcane genes, will revolutionize the application of biotechnology to crop improvement. Genetically modified sugarcane with increased resistance to biotic and abiotic stresses, yield and juice could become useful in breeding for better varieties. This review outlines some of the biotechnological developments that are in place and tailored to address important issues related to sugarcane improvement.

Keywords: Sugarcane, Biotechnology, *in vitro* culture, Mutagenesis, Transgenic plants Genomics, Stress tolerance, Molecular markers

[*] Email: rm.devarumath@vsisugar.org.in, rdevarumath@gmail.com.

INTRODUCTION

Sugarcane is cultivated in the tropical and sub-tropical regions of more than 90 countries with area under cultivation close to 20 millions of hectares (FAO; http://apps.fao.org, http://www.illovo.co.za/worldofsugar). The Indian sugar industry plays an important role in the global market as the world's second largest producer after Brazil, producing nearly 15% sugar and 25% sugarcane per annum under a wide range of agro-climatic conditions. Currently, the industry produces around 300-350 million tonnes (Mt) cane, 20-22 Mt white sugar and 6-8 Mt jaggery and khandsari to meet the domestic consumption of sweeteners [1]. In Maharashtra, the total sugarcane cultivation is about 10.22 lakh hectares with an average yield 84.9 tonnes per hectare and sugar production of about 86.7 lakh tonnes [2]. However, sugarcane production offers continuing challenges to the development of high sugar, high yielding, abiotic and biotic resistance clones in Maharashtra. Attempts to evolve productive varieties of sugarcane are being made by conventional breeding methods for last several decades.

The progress of economically important sugarcane research emerges from the conventional breeding, genome understanding, gene discovery and molecular breeding. Sugarcane improvement, from selection of existing variation in pre-historic time to the current bi/multi-parental crossing and subsequent use of non-conventional techniques, has concentrated mostly on improving the yield and sugar content.

Cultivated sugarcane (*Saccharum* spp. hybrids, 2n=100–130) belongs to the genus *Saccharum* of the family *Poaceae.* (tribe Andropogoneae). The genus is characterized by clonal propagation, complex aneu-polyploidy and high levels of heterozygosity. It is comprised of six species, namely *S. officinarum* L. (2n = 80), two wild species *S. robustum* Brandes and Jeswiet ex Grassl (2n = 60–80) and *S. spontaneum* L. (2n = 40–128) and three secondary species *S. barberi* Jeswiet. (2n = 81–124), *S. sinense* Roxb. (2n =111–120) and *S. edule* Hassk. (2n = 60, 70, 80) [3], *S. officinarum*, *S. spontaneum* and *S. robustum* represent the basic species. *Saccharum officinarum*, however, is believed to have evolved through hybridization of species such as *Erianthus arundinaceus* (Retz.) Jeswiet, *S. spontaneum*, and *S. robustum* [4], whereas *S. barberi* and *S. sinense* are the secondary ones believed to be natural hybrids between *S. officinarum* and *S. spontaneum* [5]. The cultivated sugarcane *Saccharum* spp. is believed to have originated from complex hybridization events (termed ''nobilization'') between *S. officinarum*, *S. barberi*, *S. sinense*, and the wild related species *S. spontaneum* [6], other genera such as *Erianthus* Michx., *Miscanthus* Anderss, *Narenga* Burkiee, and *Sclerostachya* (Hack.) A. Camus is closely related to *Saccharum*. Mukherjee [7] coined the term '*Saccharum* complex' to encompass all the above species and genera constitute an inbreeding group called *Saccharum* complex [4]. The mutual relationship and actual contribution of these different genera, however, remain unclear due to their high and variably ploidy levels [8]. Current sugarcane cultivars are estimated to possess 80-90 % of the genome from *S. officinarum* and 10-20 % from *S. spontaneum* [9].

There is increasing pressure worldwide to enhance the productivity of sugarcane in order to sustain profitable sugar industries. The conventional breeding programmes are being run successfully at different sugarcane research institutes to develop new hybrid varieties with high yielding potential and high sugar contents. A series of backcrosses to *S. officinarum* resulted in cultivars with higher yields, improved ratooning ability and disease resistance.

However, conventional breeding required for developing new varieties as high as 12 to 15 years to develop and release an elite sugarcane variety. It also allows the perpetuation of diseases from generation to generation. Thus, new approaches in plant biotechnology have opened up numerous opportunities that can be applied for precise breeding to improve varieties for specific objectives and also for quick multiplication of new varieties. Major biotechnological challenges and opportunities lie in improving sugarcane productivity.

Recent advances in biotechnology offer several opportunities to address issues related to the development of novel and high-yielding cultivars. Attempts are being made to use recent biotechnological innovations to tackle the challenges facing the current plant improvement programmes. Slow multiplication rate and rapid spread of newly released sugarcane varieties are other major constraints. Due to limited availability of seed cane of a new variety at the time of its release, it further takes about 8-10 years to cover the desired area for commercial cultivation. By this time, the variety starts deteriorating due to biotic and abiotic stresses. It has been realized that the conventional methods of plant multiplication are unable to meet the growing demand of seed cane material of newly released varieties. Therefore, exploring the biotechnological tools is quite essential for fast multiplication of newly released varieties. This article describes the biotechnological approaches for use in sugarcane improvement at Vasantdada Sugar Institute, Manjari-Pune, India.

Broadening Genetic Variability through Somaclonal Variation and in Vitro Mutagenesis

Micropropagation is an *in vitro* method for clonal multiplication of plants using meristematic cells/tissue as the explants. Sugarcane plants propagated *in vitro* from meristems are considered to be more genetically and phenotypically stable compared to those produced from callus. Genetic variability has been reported in tissue-cultured sugarcane. *In vitro* culture-induced variability, although infrequently beneficial, is undesirable for both commercial propagation and germplasm storage.

Devarumath et al. [10] made a comparative study to evaluate the field performance of meristem culture plantlets and conventional setts as seed source of two commercially popular sugarcane varieties, viz., Co 86032 and CoC 671 for various agro-morphological and quality traits. RAPD technique was used to assess the genetic fidelity of the meristem culture plants in relation to the mother plants. Plants derived from the meristem culture did not differ in the key agro-morphological traits from the plants raised by conventional setts. The molecular characterization of sugarcane using 31 RAPD primers showed that the amplification products were monomorphic across all micropropagated plants along with donor parent.

Sugarcane varieties Co 94012 [11] and VSI 434 are the new sugarcane varieties obtained through somaclonal variation from CoC 671. The genetic variation of micropropagated plantlets was assessed by RAPD markers. The banding pattern of PCR amplified products from micropropagated plantlets showed that most of them were monomorphic in both the varieties. The amplification pattern of VSI 434 and Co 94012 with primers OPA 17 and OPA 19 differed from parent CoC 671, respectively [12].

Sugarcane somaclones derived by callus culture of the sugarcane variety CoC 671 were evaluated for their quantitative attributes and assessed for the genetic variation by RAPD analysis. Field evaluation studies 12th month, VSI 2179 gave higher cane yield compared to parent. VSI 1748 and VSI 2179 gave significantly higher sugar yield and VSI 1748, VSI 2003 and VSI 2179 were significantly superior for brix, sucrose and CCS percentage over their parent CoC 671. RAPD profiling of somaclones, VSI 1733 with 21 primers (13.8%), VSI 1748 and VSI 1855 with four primers (3.0%) and VSI 2003 and VSI 2179 with one primer (0.3%) revealed polymorphism compared to CoC 671. These promising clones are being evaluated in different agro-climatic zones of Maharashtra [13].

Tissue culture induced variation ("somaclonal variation") may offer additional variation to that induced through mutagenesis and such a variation can be most effective if it is successfully associated with cellular level selection and handling of large populations for screening [14]. Physical and chemical mutagens have been applied to *in vitro* cultures so as to enhance the frequency of genetic variation and obtain beneficial modifications in the cultivars. We have been working towards employing *in vitro* culture combined with EMS and radiation induced mutagenesis in the improvement of sugarcane [15, 16]. Earlier studies using EMS induced mutagenesis in two sugarcane clones viz., CoC 671 and VSI 434 were used for induction of genetic variability through *in vitro* mutagenesis using chemical mutagen ethyl methane sulphonate (EMS). Apical meristematic region was used for callus induction on 4.0 mg/l 2,4-D. Actively growing callus was treated with three different doses of EMS (0.0, 0.1, 0.3 and 0.5%) with treatment of 1, 3 and 5 hours. Maximum callus proliferation and plantlets regeneration was observed in control and minimum at 0.3% EMS. The regenerated plantlets were hardened and planted in the field. Three mutants TC 2513 (derived from CoC 671), TC 2543 and TC 2556 (derived from VSI 434) differed from each other as well as from their donor as evidenced from morphological as well as qualitative characters selected for clonal trial in the breeding programme [15].

Dalvi et al. [17] studied the genetic improvement of sugarcane variety CoC 671 which was carried out through somaclonal variation using 0.8M EMS containing medium. All the regenerated plants were hardened in green house plant in the field. On the basis of biometric and biochemical parameters, these plantlets were evaluated in the rod row trial and then for smut resistance supplementing with smut inoculums and then analyzed by PCR. Two clones were promising: TC 906 (resistant to smut) and TC 922 (moderately resistant to smut). These two clones were superior in single cane weight, cane yield and diameter over the parent CoC 671 showing phenotypic variation.

Somaclonal variation in combination with *in vitro* mutagenesis can be beneficial for the isolation of salinity tolerant lines in a short duration employing *in vitro* selection. This project was carried out under the DAE-BRNS collaborative research with Nuclear Agriculture and Biotechnology Division, BARC, Mumbai. Earlier studies were based on using radiation induced mutagenesis and *in vitro* development of salt selection mutants in sugarcane using 10, 20, 30, 40, 50, and 60 Gy gamma-ray irradiated cultures. In our studies, we have used popular sugarcane varieties Co 86032, Co 740 and CoM 0265 with *in vitro* mutagenesis in combination with cellular selection for salt tolerance (Fig. 1a-1c).

Nikam et al. [16] employed *in vitro* mutagenesis for the selection of salt tolerance in sugarcane cv. Co 86032 using embryogenic callus. Sugarcane leaf base segments were cultured on MS medium with 3 mg l^{-1} 2, 4-D for 4 weeks in dark. Embryogenic callus cultures were subjected to different doses of gamma radiation (10 to 80 Gy). The 20 Gy consider as LD_{50} irradiated cultures exhibited almost 50% survival response. To evaluate the salt tolerant lines, the embryogenic calli were exposed to different levels of NaCl (50 to 250 mM). Irradiated and non-irradiated cultures showed a decrease in the callus growth with increasing selection pressure of salt in terms of relative growth rate (RGR). Higher amounts of free proline, glycine betaine and MDA were accumulated in NaCl stressed calli. The Na^+ content increased and K^+ content decreased with increasing levels of NaCl. This mechanism implies that sugarcane may be considered as a Na^+ -excluder. The accumulation of salt ions and osmolytes may play an important role in osmotic balance in the sugarcane cells under salt stress. A similar approach was employed in the sugarcane variety Co 740 calli. A total of 214 salt selected plants were grown to maturity and the agronomic performance of mutant clones was evaluated under normal and saline conditions. The 24 clones were characterized for biochemical attributes related to salt stress and showed better agronomic performance in terms of Brix%, number of millable canes, girth and yield [18].

Tissue culture plantlets can be used for screening salt tolerance in sugarcane as shown by Karpe et al. [19]. A comparative study was made to assess salt stress responses of sugarcane (*Saccharum officinarum L.*) var. CoC 671 and Co 86032 using *in vitro* plantlets by subjecting them to increasing concentrations of NaCl (0, 50, 100, 150, 200 and 250 mM) and checking relative growth rate (RGR), membrane damage rate (MDR), soluble proteins, osmolytes (proline, glycine betaine), ions (Na^+ and $K^{+)}$ and activity antioxidant enzymes (peroxidase, ascorbate peroxidase, guaiacol peroxidase, catalase and superoxide dismutase). As the concentration of NaCl increased, the RGR was found to decrease by 42.1 and 77.7%, the MDA level increased by 32.5 and 55.8% and an increase in proline of about 43 and 189% was seen in CoC 671 and Co 86032 respectively. CoC 671 was adapted to higher Na+ concentration (150 mM) than Co 86032. For the K+ accumulation, it displayed similar patterns as in Na+ accumulation. In general, it was observed that in all cases except catalase, CoC 671 displayed higher tolerance to NaCl (up to 150 mM) than Co 86032 (up to 100 mM). Based on these results, it is suggested that CoC 671 displayed NaCl tolerance up to about 150 mM, while that of Co 86032 was around 100 mM.

Figure 1a. (A-E) Callus induction, regeneration and plant establishment.

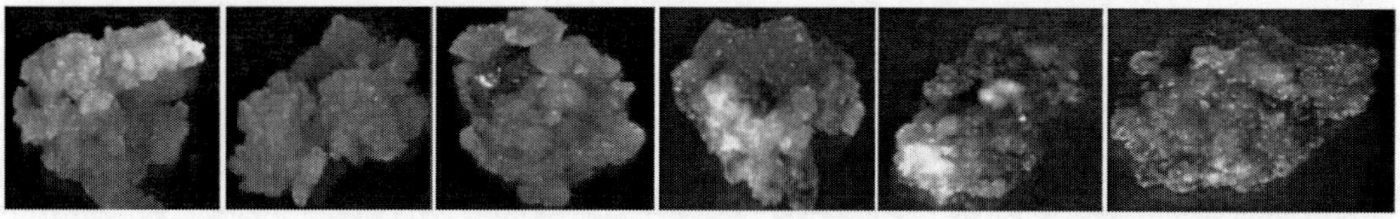

Figure 1b. Effect of ã- radiation (20 Gy) and NaCl (00, 50 to 250 mM) on callus after 30 days of culture.

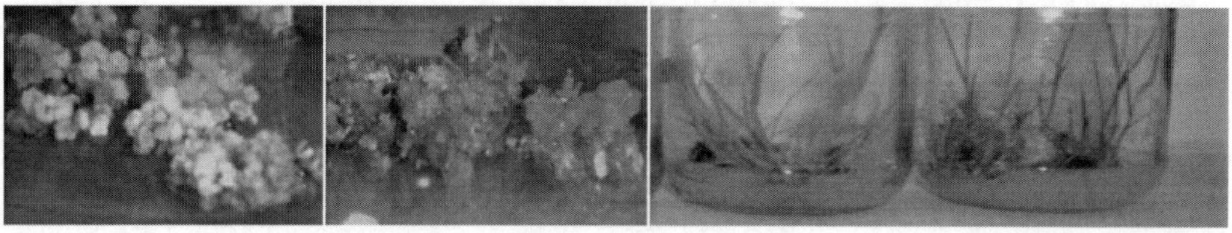

Figure 1c. Plant regeneration from callus cultures after radiation exposure (20 Gy) and selection on salt medium.

Field Evaluation

The field evaluation of both radiation treated plantlets and radiation with salt selected plantlets were transplanted in to the field for preliminary screening and scoring morphological variations. Twelve-month-old somaclones were selected on the basis of brix%, cane diameter, number of millable cane and morphological variation such as change in cane colour, canopy structure, tillering, waxiness, plant habit and bud shape in ground nursery. Further, selected promising clones from ground nursery with the parent variety were assessed for their agronomic performance such as germination, number of tillers per plant, number of millable canes per stool, girth, leaf length, leaf breadth and cane Brix% were evaluated at the 10th and 12th month after planting. From ground nursery, best performing clones that recorded ≥ 21 % Brix were selected and planted in clonal trial I in augmented field (Rod row) design evaluated for agronomic performance with parent. From this clonal trial I selected clones were further evaluated in clonal trial II in randomized block design (RBD), with three replications along with parent and standard checks. All the biometric and biochemical parameters viz. total height of cane, diameter of cane, single weight of cane, brix%, pol% and commercial cane sugar (CCS%) were recorded in 10th and 12th month and compared with parent and standard check varieties. Randomly three canes from each row were taken for analysis. Further research is in progress for molecular characterization using DNA markers and promising clones are being evaluated in saline soil condition (data not shown).

NUCLEAR AND PLASTID TRANSFORMATION IN SUGARCANE

Traditional back crossing to recover elite genotypes with desired agronomic traits is a very difficult task in sugarcane due to its complex polyploid nature, variable fertility and genotype versus environment interactions. This makes sugarcane an ideal candidate for genetic engineering. The availability of tissue culture regeneration system from various explants makes this crop a suitable candidate for genetic manipulation. In addition, the gene transfer techniques are well established in sugarcane and the vegetative propagative nature of sugarcane can easily pass the transgene to progenies and maintain the same without loss. Tremendous progress has been made in sugarcane genetic engineering and several genes targeted towards sugarcane improvement have been introduced into sugarcane. Genes for disease/pest resistance, for drought tolerance, and for quality improvement such as sugar accumulation have been introduced into sugarcane [14].

The success of transgenic sugarcane plant production depends on three major aspects: the gene transfer technique used for transformation, availability of highly regenerable the target tissue/explants and an efficient selection system for the screening of transformed tissue from the non-transformed ones. Somatic cells with good embryogenic potential are ideal targets for integration of transgenes since each somatic embryo has the potential to become an individual plant. Various explants types (axillary buds, apical meristems, immature inflorescences, leaf segments) have been used successfully to regenerate full plants in sugarcane indicating that a wide range of totipotent target tissues are available for genetic transformation. Several reports on nuclear transformation of sugarcane are available, which involve different methods for transformation like, particle bombardment, *Agrobacterium*-mediated method [20] and electroporation [21]. We reported *Agrobacterium* mediated transformation to produce transgenic sugarcane for borer resistance was using *Cry* 1Aa3 gene [22].

Recent advances in the field of genetic engineering include targeting the chloroplast genome for expression of foreign genes [23, 24, 25, 26, 27] If the plastid transformation technology is developed for sugarcane, development of genetically modified plants that have transgene containment and also have a higher expression of foreign protein will be possible, which may aid in resistance management (herbicide tolerance, insect resistance) [28], if these genes are transformed into the plastid genome.

The fully sequenced chloroplast genome, efficient regeneration system and the possibility to conduct regeneration rounds in sugarcane makes it the model organism for plastid transformation in monocots, which is not yet available. It may also allow us to identify a useful selection system for monocot plastid transformation, in general. With this view, we attempt to develop the plastid transformation technology for sugarcane.

We attempted the plastid transformation in sugarcane in collaboration with Prof. Dr. Ralph Bock, (Max Planck Institute of Molecular Plant Physiology, Germany) under the DAAD (German Academic Exchange Service) funded Sandwich Model Fellowship Programme. In Germany, the plastid transformation vector with the *nptII* gene, Prrn promoter and TrbcL terminator developed by Prof. Dr. Ralph Bock's group at MPI-MP (Germany) was used for plastid transformation of sugarcane [29, 30].

The young leaf tops of sugarcane were harvested from 6-8 month old plants growing under greenhouse conditions. The leaf rolls were dissected out in the laboratory, sterilized and the outer leaf scales were removed under aseptic conditions. The leaf rolls were cut into transverse sections and such leaf roll discs were placed on callus induction medium and incubated at 28°C in dark. Callus formation was observed in these explants after three to four weeks which later on produced shoots when placed on shoot induction medium. Only the embryogenic type of callus was selected for the transformation experiments.

Transgenic plants can only be regenerated from cells competent for both regeneration and integrative transformation [31, 32]. Availability of target tissue competent for regeneration therefore becomes an essential requirement of a gene transfer system for production of transgenic plants [31]. In the present work, three genotypes namely Q117 (Australia), NCO310 (Australia) and NA85-1602 (Argentina) were tested for their regeneration efficiency. Also the regeneration efficiency of these cultivars was compared to four commercial Indian genotypes (CoC 671, Co 86032, CoVSI 9805, VSI 434) imported from Vasantdada Sugar Institute, India. The genotypes Q117 and NA85-1602 were found to have higher regeneration efficiency than the other genotypes. These genotypes were included in the further work on transformation.

Two different types of gene guns, namely the PDS-1000/He system and the Particle Inflow Gun (PIG), are available for the particle bombardment-mediated delivery of the foreign DNA into plant cells. Also, the DNA can be coated on using gold or tungsten particles. The PDS-1000/He system is available with mono and hepta adaptors. In order to test the transformation efficiency resulting from bombardment with different gene guns, particles and adaptors, the leaf roll discs of sugarcane were bombarded using the vector pAHC25 (*gus* gene containing vector), making use of different combinations of the guns and particles and adaptors tested on three genotypes of sugarcane (Q117, NA85-1602 and NCO310). A transient GUS assay was done after 48 hours of bombardment. No significant difference was obtained when the number of blue loci on different explants was counted. In all the further experiments, both the gene guns and particle types were used, while only making use of the mono adaptor with the PDS-1000/He system. Double and triple shots on the same explants were also included in the transformation experiments.

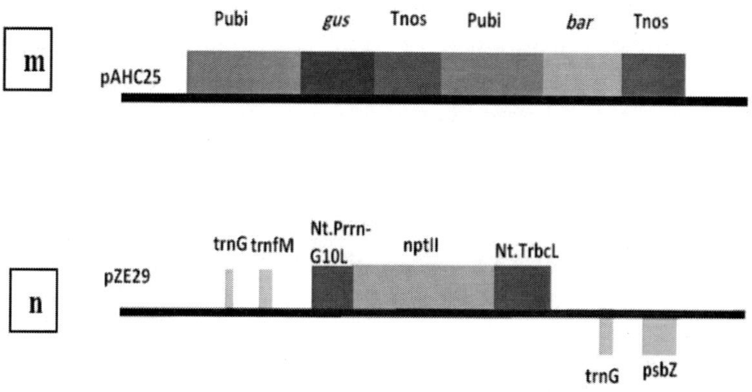

Figure 2 a-n: Representation of genetic transformation in sugarcane.

a-f: A leaf based regeneration system for sugarcane: a- Seven months old sugarcane plants in greenhouse; b- Top of sugarcane plant, containing shoot apical meristem; c- Leaf rolls dissected and transverse sections prepared under aseptic conditions; d- Leaf roll discs placed on callus induction; e- Shoot formation on shoot induction medium (upper half of picture)/Embryogenic callus formation on callus (lower half of picture) f- Multiple shoots obtained from regenerating plant; g-f: Gene guns used for transformation: g- Biolistic gene gun (PDS/He 1000); h- Particle inflow gun; i- leaf roll disc explants on osmotic medium before bombardment; j- Transient GUS expression seen in the leaf roll disc explants; k- Shoot regeneration in control (wild-type explants on regeneration medium without antibiotic) plate; l- transformed explants on selection (regeneration medium with geneticin 75 mg/L); m-nuclear transformation vector pAHC25 used for studying transient GUS expression in sugarcane. pAHC25 contains the uidA (gus, β-glucuronidase) and bar (bialaphos resistance)genes under the control of the maize ubiquitin (ubi1) promoter and its intron (ubiII), followed by the nos terminator; n-plastid transformation vector pZE29 with nptII gene conferring resistance to geneticin with the tobacco Prrn-G10L promoter and TrbcL terminator while trnG, trnfM, trnG and psbZ are the flanking sequences indicating the intergenic region between trnfM and trnG as the insertion site of the transgene into the plastid genome of sugarcane.

After 7-21 days of incubation on callus induction medium, the leaf roll discs showing good *in vitro* response (in terms of pro-embryogenic formation) were selected for transformation. Plastid transformation experiments with the vector pZE29 (containing *nptII* gene) were performed. We used nuclear transformation experiments with vector (pUBInptII) as positive control for the tissue culture regeneration and transformation process. The transformed explants were placed on antibiotic selection medium containing geneticin (50-70 mg/L). The medium was changed every two weeks. Most of the regenerating plants did not survive after repeated sub-cultures on selection medium. The nuclear transformation (positive control) with plasmid pUBInptII gave resistant lines and the PCR analysis of these lines showed the presence of the *nptII* gene.

It was observed that the leaf roll discs showed good response in terms of embryogenic callus formation and regeneration, during the initial cycles of selection on antibiotic selection medium, but the rate of regeneration decreased gradually. With the callus system, the regeneration was very slow during the initial cycles of selection, but the regenerating plants obtained were relatively stable in their response even after prolonged exposure to higher concentrations of antibiotic.

To enrich the transplastomic genome in the primary transformed shoot, additional regeneration rounds are necessary. Embryogenic callus induction and regeneration was obtained from sugarcane plants in tissue culture. This technique would later be adopted for the plastid-transformed lines of sugarcane. The work is in progress (data not shown). (See Fig. 2 for an overview of the different steps in tissue culture and transformation of sugarcane followed in the present work).

In DBT funded project plastid transformation was done using the sugarcane cultivar Co 86032 and CoC 671 at VSI. The tissue culture protocol for embryogenic callus induction and antibiotic sensitivity test for geneticin was optimized for the selection of putative transplastomic plants [33, 34, 35].

Genomics for Salinity and Drought Stress Tolerance

Salinity and drought are the major environmental factors that limit crop productivity mainly due to alterations in water relations, ionic, and metabolic perturbations; generation of reactive oxygen species (ROS); and tissue damage. FAO estimates suggest that around 6% of the world's total land area and 20% of irrigated land are affected by high salinity.

Analysis of the expression of these stress-responsive genes in sugarcane plants that were under water deficit stress revealed a different transcriptional profile to that which correlated with sucrose accumulation. Prabu et al. [36] identified differentially expressed transcripts in response to water deficiency stress in sugarcane cv. Co740 using PCR-based cDNA suppression subtractive hybridization technique. The EST database generated showed 62% similarity with known functional genes, 12% with hypothetical proteins of plant origin, while 26% represented new unknown sequences.

In an attempt to understand the molecular basis of salt-stress response in sugarcane, physio-biochemical assays and cDNA-RAPD-based gene expression studied under high salt (2% NaCl) stress response, Pagariya et al. [37] carried out cDNA-RAPD-based gene expression at early growth stage in tolerant sugarcane variety Co 62175. The comparative rates of total protein, proline content and lipid peroxidation were found steadily increased, while total chlorophyll content was decreased in leaves of salt treated over untreated sugarcane plants at corresponding increase in soil electrical conductivity. The comparative transcript responses to salt stress were monitored by ribotyping of both treated and untreated sugarcane plants at early growth stage. Among 335 differentially expressed transcript-derived fragments, 156 up- and 85 down-regulated were re-amplified and sequenced. Further, sugarcane responses at germination, tillering and respective recovery stages to high salinity at physio-biochemical and molecular level. Our results indicated that there was a negative relationship between catalase, and peroxidase activity with lipid peroxidation and SOD activity. Increase in MDA and SOD levels at the earlier stages of stress and later increase in CAT and POD levels on prolonged stress was evident. Thus they can be used as indicators of stress for sugarcane plants facing unfavorable environmental conditions. At molecular level, we have identified 137 salinity tolerant candidate cDNAs from sugarcane by cDNA-SSH, representing 20% of which are novel sugarcane genes. [38, 39].

Genomics for Biotic Stress Tolerance

In another study cDNA-SSH library was constructed and analyzed to identify the up-regulated genes in sugarcane under SCGS infection condition. Subtracted library highly represented genes potentially involved in cell rescue, defense, ageing and apoptosis (13.1%). The forward SSH approach implemented, allowed to explicate the transcriptional regulatory mechanisms of sugarcane in response to SCGS infection and isolated the *R2R3-MYB (SoMYB*18) gene, a potential candidate playing important roles in the regulation of secondary metabolism, signal transduction during biotic, abiotic and other environmental stresses [40]. In an attempt towards studying the host-pathogen interaction to decipher the molecular basis of virulence of sugarcane SCGS disease, Kawar et al. [41] isolated partial genome of first Asiatic strain of phytoplasma (SCGS) by genomic-SSH. The library yielded 83 SCGS specific fragments representing approximately 42% of the chromosome of Sugarcane grassy shoot phytoplasma, comprising approximately 85 predicted partial phytoplasma CDS. Further, a species specific detection method was developed for early detection of SCGS infection [42].

MYB Transcription Factors

Plant R2R3 *MYB* transcription factors play an important role in various plant-specific regulatory processes. A large group of transcription factors have been classified into bZIP, MYB, WRKY, AP2/DREBP and some zinc finger like proteins as a part of gene regulation in response to biotic and abiotic stress conditions. The *MYB* family consists of a conserved DNA-binding domain, *MYB* domain, which consists of 1-3 imperfect helix-turn-helix repeats (R1, R2 and R3); animal MYBs containing three repeats and plants two repeats R2 and R3. The *MYB* genes constitute the largest gene family in plants, with over 126 R2R3-MYB members in *Arabidopsis*, 109 in rice and 80 in maize. This *MYB* gene family has been of immense importance in transcriptional control studies in higher plants due to its key role in the regulatory networks like development, metabolism and responses to biotic and abiotic stresses.

The sugarcane (*Saccharum officinarum*) stress related *MYB* transcription factor gene, ScMYBAS1, demonstrated induced response to the water deficit and salt stress [43]. To elucidate its stress tolerance mechanism at the transcriptional level, the promoter (PScMYBAS1, 1,033 bp) flanking the 50 ScMYBAS1 coding region from the sugarcane genome was isolated and characterized. A series of PScMYBAS1 deletion derivatives from the transcription start site (-56, -152, -303, -442, -613, -777, -843, -1,033) were fused to the *uidA* reporter gene (GUS) and each deletion construct was analyzed by *Agrobacterium*-mediated transient transformation in tobacco leaves subjected to dehydration, salinity, cold, wounding, gibberellic acid (GA), salicylic acid (SA), and methyl jasmonic acid (MeJA). Deletion analysis of the promoter, PScMYBAS1, suggested that the 303-bp promoter region was required for basal expression. Promoter fragments, 777 bp or longer showed two-fold to four-fold increased induction of GUS in response to abiotic stress (dehydration, salt, cold, wounding) and hormone (SA, MeJA) treatments. These findings further throw a light on our understanding of the regulation of ScMYBAS1 expression and provide a new stress-inducible promoter system in transgenic plants.

To enhance the insight of the response to the changing environments and the role of *MYB* genes in the control of plant-specific processes, isolation and analysis of *MYB* genes from the wild relative species of sugarcane was undertaken. To cope up with the upcoming research findings and to unravel the structural predictions of transcription factor proteins, computational sciences have been proposed. Many online protein structure predicting servers available at ease viz; ProSA, PROMOTIF, PROCHECK, I-TASSER, ProFunc, Verify3D; were used for conducting these experiments.

Primers were designed using the conserved sequences of MYB transcription factor among several genes sequences with the highest similarity to *Saccharum officinarum* hybrid (Accession No: FJ560976). The PCR product obtained was purified and cloned into T/A vector according to manufacturer's instructions and sequenced and partial sequences of *MYB18* were submitted to EMBL database from different species of sugarcane and related genera as *Saccharum officinarum* (Vellai, 1704 bp), *Saccharum robustum* (1702 bp), *Saccharum barberi* (Pathri, 1694 bp), *Erianthus ciliaris* (1691 bp), *Erianthus elegans* 1679 bp, *Erianthus arundinaceus* (1692 bp) *Narenga* (1673 bp) with EMBL accession No. HF546401 to HF546407 respectively. Other partial sequences of MYB 21 were submitted to EMBL database *Saccharum officinarum* (Vellai, 534 bp), *Saccharum barberi* (Pathri, 529 bp), *Erianthus arundinaceus* (526 bp), *Saccharum officinarum* (545 bp) with EMBL accession No. HF546408 to HF546411 respectively.

In silico analysis of the four *MYB* genes isolated from *S. officinarum* (Vellai), *S. barberi* (Pathri), *Saccharum spontaneum* and *Erianthus arundinaceus* sugarcane species and wild related genera. The exon and intron pattern of the four *MYB* genes was deduced using fGENESH Hidden Markov Model (HMM) structure prediction tool (www.softberry.com/berry.phtml). Deduced amino acid sequence was used to predict the protein domains with position-specific iterated BLAST (PSI-BLAST) for getting the homologues. Sequence alignment and phylogenetic tree analysis were performed by using EBI ClustalW2 and MEGA 4.0 respectively. The secondary and 3D structure prediction of four *MYB* genes was done by the iterative threading assembly refinement, I-TASSER server and *Ab initio* modeling). Validation and inspection of the four *MYB* genes was carried out using PROCHECK, ProSA, PFAM and Q-site finder tool.

The full-length sequences of SoMYB18, SbMYB18, EaMYB2R and SsMYB2R amplified using the specific set of MYB primers showed higher identities with the available full-length MYB sequences from NCBI database; *Saccharum officinarum* hybrid FJ560976.1, *Zea mays* NM_001138598.1, *Sorghum bicolor* XM_002440514.1, *Oryza sativa* OsMYB16 AJ495784.1 and OsMYB18 AJ495786.1. The SANT/MYB DNA binding domains are located at the N-terminal of the deduced protein with ~50 amino acids motif consisting of the R2 and R3 regions. The highly conserved tryptophan (W) residues are seem to be involved in the folding of the DNA binding domain are denoted in our four *MYB* sequences (Fig. 3). To understand the relationships among the R2R3 domains from several MYB-related proteins, a dendrogram was raised. It showed relationship with both dicot and monocot homologues suggesting that they may be evolved through same ancestors of the R2 and R3 family.

The I-TASSER server generated five 3D atomic models from multiple threading alignments and I-TASSER for SoMYB18, SbMYB18, EaMYB2R and SsMYB2R gene of which the model showing the maximum C-score, TM-score, RMSD (Root Mean Square

Deviation), number of decoys (>600) and cluster density was selected for further internal evaluation of self-consistency checks. The final tertiary structures of each *MYB* genes were generated using the Pymol Molecular Graphics System (http://pymol.org/ep). The tertiary structures showed presence of beta-sheets only in SoMYB18 [44] and SsMYB2R [45] in contrast to SbMYB18 [44] and EaMYB2R [45] genes thus suggesting the evolutionary aspect of the *MYB* transcription factor family in sugarcane and its relative species. The relatively low percentage of residues in the disallowed regions of Ramachandran plot suggested the acceptable quality of the four *MYB* gene models (Fig. 3a and 3b represents the bioinformatic analysis of the *MYB* genes).

The common attribute of the R2R3-MYB and MYB-related proteins is the wide diversity of functions to sustain under stress conditions. Though the task of understanding whole of the *MYB* family genes from plant genomes is intricate and challenging, a small bioinformatical effort putforth in this study might provide robust foundation for predicting the protein-protein and/or protein-DNA interactions of *MYB* genes in further studies. The secondary and tertiary structures of *MYB* genes established important information regarding the DNA binding domains helpful for activating other genes and its expression studies under stress conditions.

Similarly Prabu and Theertha Prasad [43] showed that sugarcane (*Saccharum officinarum*) stress-related *MYB* transcription factor gene *ScMYBAS1-3* elucidate its sequence-to-structure-to-function paradigm, the putative three-dimensional structure of ScMYBAS1 was generated using threading assembly refinement (I-TASSER) server. Further, PROCHECK, Verify-3D, PROMOTIF and ProSA programs were used to test the quality of model and the scores were within the recommended intervals. The models shed valuable information necessary for future identification of DNA binding regions and the prediction of co-regulated stress induced genes by docking studies.

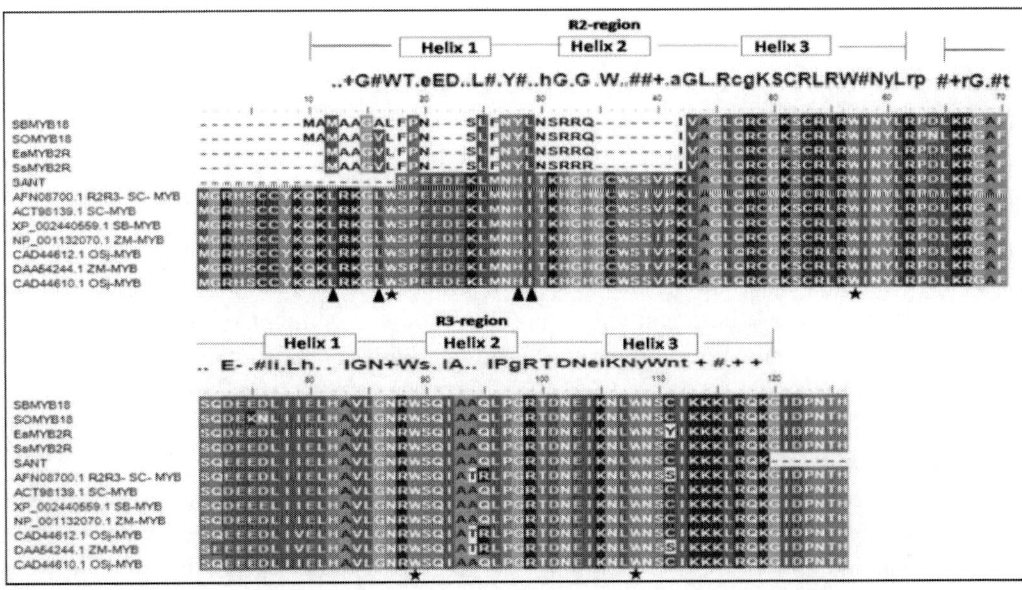

Figure 3a. Sequence alignment of SbMYB18, SoMYB18, EaMYB2R and SsMYB2R based on consensus R2R3 SANT/MYB DNA-binding domain. SB- *Saccharum barberi* Pathri, SO- *Saccharum officinarum* vellai, Ea- *Erianthus arundinaceus*, Ss- *Saccharum spontaneum*, SANT- Domain, SC- *Saccharum officinarum* hybrid, SB- *Sorghum bicolor*, ZM - *Zea mays* OS- *Oryza sativa*.

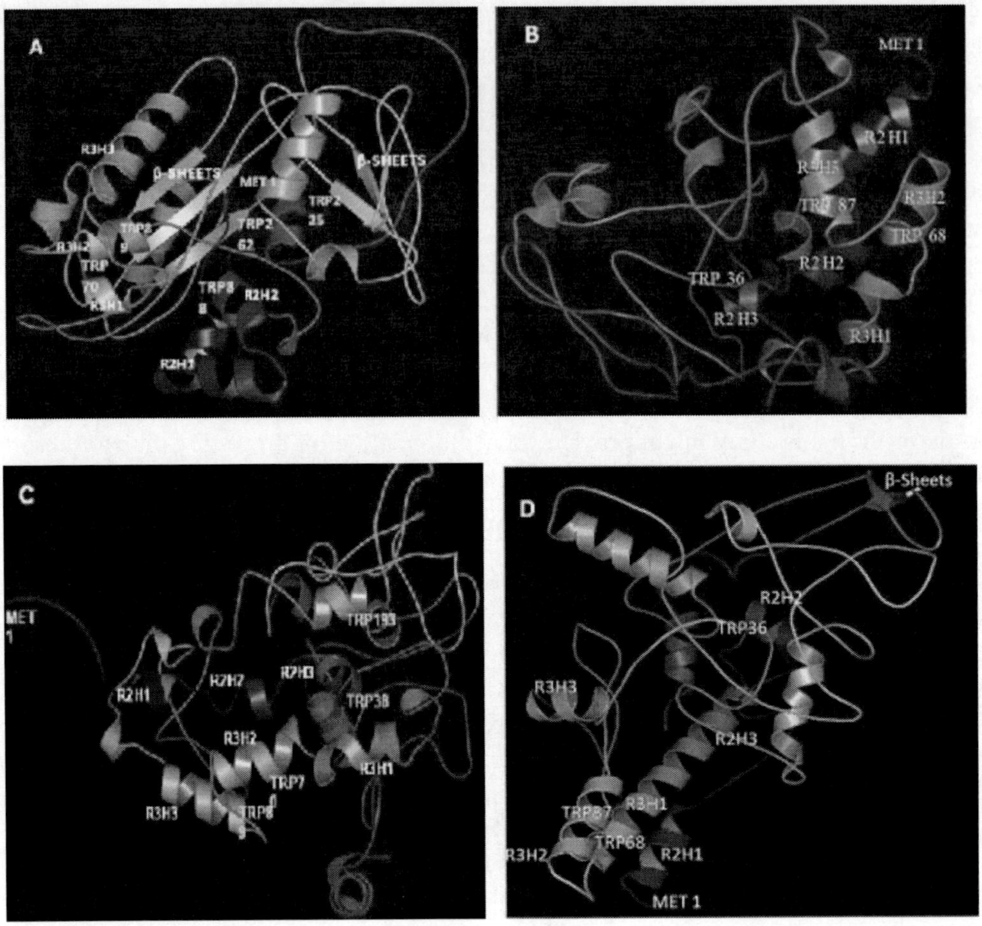

Figure 3b. Predicted 3D models for A- SoMYB18, B-EaMYB2R, C-SbMYB18 and D-SsMYB2R.

Use of Molecular DNA Markers in Sugarcane

Molecular markers serve as an excellent genetic diagnostic tool to analyze large genomes. The markers available now are either restriction endonuclease or polymerase chain reaction (PCR) based, or a combination of both. Although a number of molecular markers have been developed, the most commonly used markers for genotypic analysis include restriction fragment length polymorphism (RFLP), random amplified polymorphic DNA (RAPD), simple sequence repeats (SSR), inter-simple sequence repeats (ISSR) and amplified fragment length polymorphism (AFLP).

Single Nucleotide Polymorphism (SNP) are point mutations in which one nucleotide is substituted at a particular locus. They represent an inexhaustible source of polymorphism which is useful in high resolution mapping studies. These can be put to use where the genomics is well advanced. Apart from these markers, several markers based on these marker systems have been developed which have also been tried in sugarcane. Expressed Sequence Tag (EST) based markers utilize the expressed portion of the genome or the cDNAs. EST sequences represent real-function genes and thus are more useful as genetic markers. A large number of ESTs have already been reported in sugarcane. The sugarcane EST project SUCEST has built a database containing 2,38,000 ESTs from 26 cDNA libraries constructed from several organs and tissues sampled at different intervals (http://sucest.lad.ic. unicamp.br/en/).

The EST database in the public domain also serves as a readily available inexpensive source of microsatellite markers. Single Strand Conformation Polymorphism (SSCP) from genomic sequences as well as ESTs has been used in sugarcane. Here, the amplified products are converted into single strands and electrophoresed. Polymorphism arising out of conformational changes in the single strand is visualized in this case. The exact molecular nature of these variations will be understood after cloning and sequencing of the individual conformers. Targeted Region Amplified Polymorphism (TRAP) is a PCR based marker system where an EST sequence is used to design primers along with an arbitrary sequence. A fixed primer is designed from an EST sequence and an arbitrary primer of the same length with an AT or GC rich motif (to anneal with an intron or exon respectively) is designed. Sequence Related Amplified Polymorphism (SRAP) has also been used in sugarcane for various purposes like mapping studies. Conserved Intron Scanning Primers (CISP) is another marker system based on the conserved sequences that has been put to use in sugarcane.

An insight into the molecular marker systems available for genotyping explains the different ways in which each of the molecular markers systems detects the polymorphism between individuals further, their efficiency, utility and limitations are also discussed. The advent of molecular markers has certainly facilitated plant genotyping which is an easier and rapid task with reasonable accuracy and resolution, to be an integral part of crop improvement programmes.

Thus, in sugarcane, DNA markers have been used to assess the available germplasm for genetic variability, for fingerprinting of the elite genetic stocks, assessing of genetic diversity, increasing the efficiency of trait selection, to construct the genome maps and to tag genes for economically important traits and for the comparative and functional genomics studies and diagnostics. Genomic DNA is a pre-requisite for the genetic diversity analysis of crop plants. We have described a simple and user-friendly protocol for extraction of DNA from dried leaf samples. This protocol does not require use of liquid nitrogen making it advantageous over other protocols available for the genomic DNA extraction from sugarcane [46].

We used in sugarcane various molecular marker systems including Random Amplified Polymorphic DNA [40], ISSR [47, 48], SSR [48], TRAP and SNP [49] to assess the genetic diversity in elite and exotic sugarcane germplasm.

Devarumath et al. [48] characterized 81 sugarcane genotypes for genetic diversity using Inter Simple Sequence Repeat (ISSR) and Single Sequence Repeat (SSR). A total of 13 ISSR primers used and produced 65 amplified fragments, of which 63 (96.5 %) were polymorphic. The Polymorphic Information Content (PIC) value ranged from 0.11 (UBC824) to 0.45 (UBC825) primers with an average value of 0.28. The primer UBC 817 and UBC 825 exhibited highest resolving power (Rp) value 3.8 among thirteen primers. Genetic similarity (GS) by Jaccard's similarity co-efficient ranged from 0.23 to 0.95 with a mean of 0.59. The PIC value ranged from 0.06 (VSICRAD4) to 0.55 (VSICRAD26) primers with an average value of 0.17. The primer VSICRAD23 exhibited highest resolving power (Rp) value 4.3 among 28 primers. The GS by Jaccard's similarity co-efficient ranged from 0.11 to 0.91 with a mean of 0.51. Dendrogram constructed using the UPGMA cluster analysis revealed low level of correlation between genetic similarities based on the pedigree and DNA profile.

More recently, Target region amplified polymorphism (TRAP), and single nucleotide polymorphism (SNP)-based markers have also been employed to check the genetic diversity in sugarcane. The genetic evaluations of 47 sugarcane genotypes were used in the analysis [49]. TRAP is a simple polymerase chain reaction (PCR)-based marker system that takes advantage of available EST database sequence information to generate polymorphic markers targeting candidate genes and informative in amplification variation in reference to tightly linked targeted gene. The method involved in designing a fixed primer of about 18 nucleotides from EST sequences or genes of interest and an arbitrary primer about the same length is designed with either an AT- or a general collection (GC)-rich motif to anneal with an intron or exon, respectively [50, 51].

Single-nucleotide polymorphism (SNP) marker system is increasingly becoming the marker of choice replacing other marker types in many species, mainly because SNPs are common in the genome, both within and between the genes and also sequencing cost is low and SNP marker analysis can be performed easily with low error rate and amenability to high-throughput analysis. Significant resources have been devoted to the development of SNPs as high-throughput markers and also to SNP discovery [52]. Extensive SNP discovery projects have been undertaken for high-throughput use in marker-assisted breeding, for population studies in different crop plants, such as maize, rice, barley, soybean, wheat and sugarcane [52, 53]. In some species, where no genome sequence is available, large-scale SNP discovery in the genes has generally relied on the sequence information in the libraries of expressed sequence tags either (ESTs) for the direct discovery or as the basis for primer design for re-sequencing. Sugarcane is not an exception to the above and since the sugarcane genome is not yet sequenced, the ESTs have been mined as a source of SNPs [53].

Devarumath et al. [49] reported a total of 23 pairs of TRAP markers which generated 925 alleles, of which 74% alleles were polymorphic. Polymorphism was generally high (>50%), ranging from 54 to 98%. The polymorphism information content (PIC) values 0.20 varied among the primer combination ranging from 0.17 in SAI + Arbi 2 to 0.31 in GL 2+ Arbi 1 with an average of 0.24. However, the Pearson correlation between PIC and power of discrimination (PD) was found to be less significant. Single-nucleotide polymorphisms were used first time for the assessment of genetic diversity among different species of *Saccharum* and cultivated sugarcane varieties. The SNPs were detected from 454 sequencing. A total of 245 SNP markers were assayed across the 47 genotypes, and 167 SNPs were found to be polymorphic. The PIC values ranged from 0.04 to 0.38 with an average of 0.21, and their respective PD varied from 0.58 to 0.04 with an average value of 0.31. The results obtained were relatively significant when compared to the other marker systems through genetic similarity and the clusters formed in different unweighted pair group method with arithmetic mean clustering dendrogram. The clustering analysis established genetic relationship in the order of *Erianthus* > *Sclerostachya* > *Narenga* > *Saccharum spontaneum* > *S. robustum* > *S. barberi* > *S. officinarum*/cultivars. These results ratify TRAP and SNP marker systems for assessing genetic diversity studies and more diversified *Erianthus* spp. can contribute substantially towards sugarcane varietal improvement through breeding with *Saccharum* spp. or hybrid cultivars. We also carried out the functional analysis of the potential enzymes (*sucrose synthase* and *sucrose phosphate synthase*) involved in sugar modulation in the high and low sugarcane cultivars and found differential expression of the genes related to sucrose accumulation and sugar transport among high and low sugarcane cultivars. These findings reinforce the selection of diverse sugarcane cultivars for the gene expression studies targeting to quantitative traits and candidate marker determination [54].

Conclusion

Sugarcane is a source of food and fuel, and biotechnology can contribute to substantially increase the utility of this crop. The successful application of biotechnological tools will require reliable and high levels of transgene expression which is stable over the next generations. Plastid transformation has progressed gradually from *Chlamydomonas reinhardtii* to model plant tobacco and recently towards other higher plants. Most of the agronomic traits targeted for engineering via plastids were established in tobacco with an aspiration that it would be implemented in crop plants. However, till date no transplastomic crop plant could be commercialized due to various technical reasons. Thus, the plastid biotechnology for crop plants, though a novel tool for crop improvement, has several challenges which need to be addressed before realizing its true potential in improving crop plants for agronomic and industrial applications.

The availability of cellular and molecular toolbox has opened up a plethora of prospects. Innovative *in vitro* culture systems have become available with potential for rapid propagation and generating novel germplasm with desirable traits. A greater understanding of the crop using functional genomics and cellular methods will accelerate understanding responses to biotic and abiotic stresses and their management. Profiling of gene expression under the conditions that affect crop yield can aid in building up an 'expression panel' for the sugarcane cultivars which should become invaluable in the target gene selection. Gene silencing is being used in the transgenic research aimed at down-regulation of endogenous genes in sugarcane. Some of the important challenges include gene discovery, transgenic and controlled transgene expression, sucrose metabolism and photosynthesis. The advances in sugarcane biotechnology could become remarkable in the coming years, both in terms of improving productivity as well as substantially increasing the value and utility of this crop.

Acknowledgments

We are thankful to Prof. Dr. Ralph Bock (Max Planck Institute of Molecular Plant Physiology, Potsdam-Golm, Germany) for sugarcane plastid transformation vectors carrying nptII and aadA genes and also for allowing Gauri Nerkar to work at MPI-MP under the DAAD fellowship programme and Dr. Stephanie Ruf for the guidance. We also thank Zouhair Elghabi for constructing the plastid transformation vectors and Claudia Hasse for technical assistance with sugarcane transformation experiments. We thank BRNS-DAE (Mumbai) and DBT (Delhi) for funding the projects related to sugarcane at Vasantdada Sugar Institute (VSI), Pune. We also thank Mr. Shivajirao Deshmukh, Director General, VSI, Manjari (Bk), Pune, India for his constant encouragement.

REFERENCES

[1] Solomon S (2011) The Indian sugar industry: an overview. *Sugar Tech.* 1(4): 255-265.

[2] Anonymous (2013) *Cooperative Sugar* 44(7): 42.

[3] Selvi A, Nair NV, Noyer JL, Singh NK, Balasundaram N, Bansal KC, Koundal KR and Mohapatra T (2006) AFLP analysis of the phenetic organization and genetic diversity in the sugarcane complex, Saccharum and Erianthus. Genet. *Resources and Crop Evol.* 53: 831-842.

[4] Daniels J, Smith P, Paton N and Williams CA. (1975) The origin of the genus *Saccharum. Sugarcane Breed. Newsl.* 36: 24-39.

[5] Daniels J and Roach BT (1987) Taxonomy and evolution in sugarcane improvement through Breeding. Heinz DJ (Ed). *Elsevier Press,* Amsterdam pp 7-84.

[6] Sreenivasan TV, Ahloowalia BS and Heinz DJ (1987). Cytogenetics. In: Heinz DJ (ed) Sugarcane improvement through breeding. *Elsevier Amsterdam* pp 211-253.

[7] Mukherjee SK (1957) Origin and distribution of *Saccharum. Bot. Gaz.* 19: 55- 61.

[8] Nair NV, Nair S, Sreenivasan TV and Mohan M (1999) Analysis of genetic diversity and phylogeny in *Saccharum* and related genera using RAPD markers. *Genet. Resour. Crop Evol.* 46: 73-79.

[9] Piperidis N, Chen JW, Deng HH, Wang LP, Jackson PA and Piperidis G (2010) Chromosomal characterization of intergeneric introgression lines of sugarcane spp × *Erianthus arundinaceus* by genomic in situ hybridization. *Genome* 53: 331-336.

[10] Devarumath RM, Doule RB, Kawar PG, Naikebawane SB and Nerkar YS (2007) Field performance and RAPD analysis to evaluate genetic fidelity of tissue culture raised plants vis-à-vis Conventional Setts Derived Plants of Sugarcane. *Sugar Tech.* 9(1): 17-22.

[11] Jalaja NC, Sreevivasan TV, Pawar SM, Bhoi PG and Garker RM (2006) Co 94012- A new sugarcane variety through somaclonal variation. *Sugar Tech* 8: 132-136.

[12] Tawar PN, Sawant RA, Dalvi SG, Nikam AA, Kawar PG and Devarumath RM (2008) An assessment of somaclonal variation in micropropagated plants of sugarcane by RAPD markers. *Sugar Tech* 10(2): 124-127.

[13] Doule RB, Kawar PG, Nerkar YS and Devarumath RM (2008) Field performance of promising somaclonal variants and RAPD analysis to assess genetic variation in sugarcane (*Saccharum* spp.). *Indian Journal of Genetic and Plant Breeding* 68(3): 301-306.

[14] Suprasanna P, Patade VY, Desai NS, Devarumath RM, Kawar PG, Pagariya MC, Ganapathi A, Manickavasagam M and Babu KH (2011) Biotechnological developments in sugarcane improvement: An overview. *Sugar Tech* 13(4): 322-335.

[15] Nikam AA, Devarumath RM, Kawar PG, Tawar PN and Sheelavantmath SS (2011) Induce mutation in sugarcane- effects of chemical mutagen EMS on commercial cane sugar and other quality traits. *Proceedings of the Biodiversity and Biotechnology for Sustainable Development. National conference at post graduate department of studies in Botany, Karnatak University Dharwad.* pp 370- 377.

[16] Nikam AA, Devarumath RM, Sonawane BV, Kawar PG, Babu H, Tawar PN, and Suprasanna P (2011) Studies on *in vitro* gamma-irrdiation-induced mutation and salinity stress tolerance in sugarcane cv. Co 86032. *Proceeding of Balanced sugar and*

energy production in developing countries: Sustainable technologies and marketing stargies, New Delhi, India, pp 583-587

[17] Dalvi SG, Vasekar VC, Yadav A, Tawar PN, Dixit GB, Threetha Prasad D and Deshmukh RB (2012) Screening of promising sugarcane somaclones for agronomic traits, and Smut resistance using PCR amplification of inter transcribed region (ITS) of *Sporisorium scitaminae*. *Sugar Tech* 14 (1), 68-75

[18] Nikam AA, Devarumath RM, Shitole MG, Ghole VS, Tawar PN, and Suprasanna P (2014) Gamma radiation, in vitro selection for salt (NaCl) tolerance, and characterization of mutants in sugarcane (*Saccharum officianrum* L.) in vitro Cell Dev Biol-Plant DOI 10.1007/s12298-014-0231-9

[19] Karpe A, Nikam AA, Chimote KP, Kalwade SB, Kawar PG, Babu H, Devarumath RM and Suprasanna P (2012) Differential responses to salinity stress of two varieties (CoC 671 and Co 86032) of sugarcane (Saccharum officinarum L.). African Journal of Biotechnology 11(37): 9028-9035.

[20] Arencibia AD, Carmona ER, Tellez P, Chan M, Yu S, Trujillo LE and Oramas P (1998) An efficient protocol for sugarcane (Saccharum spp. L.) transformation mediated by Agrobacterium tumaefaciens. *Transgenic Research* **7**: 213-222.

[21] Downing KJ, Leslie G and Thomson JA (2000) Biocontrol of the sugarcane borer Eldana saccharina by expression of the *Bacillus thuringiensis* cry1Ac7 and *Serratia marcescens* chiA genes in sugarcane-associated bacteria. *Appl. Environ. Microbiol.* 66: 2804-2810

[22] Kalunke RM, Kolge AM, Babu KH, and Theertha Prasad D (2009) Agrobacterium mediated transformation of sugarcane for borer resistance using Cry 1Aa3 gene and one-step regeneration of transgenic plants. *Sugar Tech* 11(4): 355–359.

[23] Ruf S, Hermann M, Berger IJ, Carrer H and Bock R (2001) Stable genetic transformation of tomato plastids and expression of foreign protein in fruit. *Nature Biotech.* 19: 870-875.

[24] Bock R (2001) Transgenic chloroplasts in basic research and plant biotechnology. *J. Mol. Biol.* 312: 425-438.

[25] Daniell H, Khan MS and Allison L (2002) Milestones in chloroplast genetic engineering: an environmentally friendly era in biotechnology. *Trends Plant Sci.* 7: 84-91.

[26] Bock R and Khan MS (2004) Taming plastids for a green future. *Trends Biotechnol.* 22: 311-318.

[27] Maliga P (2004) Plastid transformation in higher plants. *Annu. Rev. Plant. Biol.* 55: 289-313.

[28] Bock R (2007) Plastid Biotechnology: Prospects for herbicide and insect resistance, metabolic engineering and molecular farming. *Curr. Opin. Biotechnol.* 18: 100-106.

[29] Nerkar G, Ruf S, Elghabi Z, Hasse C, Devarumath R and Bock R (2011) Towards plastid transformation in sugarcane. Poster presented at: Botanikertagung 2011 (18th-23rd September 2011); Frei Universität, Berlin, Germany.

[30] Nerkar G, Ruf S, Elghabi Z, Hasse C, Devarumath R, Babu H and Bock R (2014) Towards plastid transformation in sugarcane. Oral presentation at: National Seminar on Recent Advances and Challenges in Sugarcane Research, NSSR-2014 (23-24 January 2014); Mysore, India.

[31] Birch RG (1997) Plant transformation: Problems and strategies for practical application. *Annu. Rev. Plant Physiol. Plant Mol. Biol.* 48: 297-326.

[32] Potrykus I (1991) Gene transfer to plants: Assessment of published approaches and results. *Annu. Rev. Plant Physiol. Plant Mol. Biol.* 42: 205-25.

[33] Wadyalkar P, Dhamangaonkar S, Nerkar S, Patil G, Prashant GK, Devarumath RM, Ghole VS and Babu KH (2011). Improved protocols for embryogenic callus induction and direct regeneration of sugarcane (*Saccharum officinarum* L.) Cv. Co86032. Proc. of 4th International Sugar Conference & Expo, New Delhi, pp.780-784.

[34] Nerkar G, Wadyalkar P, Devarumath RM and Harinath Babu K (2010). Chloroplast transformation in sugarcane by particle bombardment Poster presented at: XXXIII Conference of Botanical Society and International Symposium on the New Horizons in Botany (10th-12th November 2010); Shivaji University, Kolhapur, India.

[35] Kale RR, Wadyalkar P, Babu KH and Devarumath RM (2013) Development of chloroplast transformation system in sugarcane (Co86032) using particle bombardment. National symposium on "Plant Tissue Culture & Biotechnology for Food and Nutritional Security" March 11-13 2013, CFTRI, Mysore (Abstract No. MP-10, P.No.55).

[36] Prabu G, Kawar PG, Pagariya MC and Theertha Prasad D (2011) Identification of water-deficit stress-upregulated genes in sugarcane. *Plant Mol Biol Rep.* 29: 291-304.

[37] Pagariya MC, Harikrshnan M, Kulkarni PA, Devarumath RM and Kawar PG (2011) Physio-biochemical analysis and transcript profiling of Saccharum officinarum L. submitted to salt stress. *Acta Physiologia Plantarum* 33(4): 1411-1424.

[38] Pagariya, MC (2012) Candidate genes as molecular markers for evaluating and validating sugarcane germplasm for salinity stress. Ph.D. Thesis, Shivaji University Kolhapur, Maharashtra, India.

[39] Pagariya MC, Devarumath R and Kawar PG (2012) Biochemical characterization and identification of differentially expressed candidate genes in salt stressed sugarcane. *Plant Sci.* 184: 1-13.

[40] Kawar PG, Devarumath RM and Nerkar Y (2009) Use of RAPD marker for assessment of genetic diversity in sugarcane cultivars. *Indian Journal of Biotechnology* 8: 67-71.

[41] Kawar PG, Pagariya MC, Dixit GB and Theertha Prasad D (2010a) Identification and Isolation of SCGS phytoplasma specific fragments by riboprofiling and development of specific diagnostic tool. *Journal of Plant Biochemistry and Biotechnology* 19: 185-194.

[42] Kawar PG, Pagariya MC, Patel SR, Dixit GB and Theertha Prasad D (2010b) An overview of partial genome sequence of First Asiatic phytoplasma strain (SCGS)-Indian Isolate. *Asian Journal of Plant Pathology* 4 (1): 16-19

[43] Prabu G and Theertha Prasad D (2012) Functional characterization of sugarcane MYB transcription factor gene promoter (PScMYBAS1) in response to abiotic stresses and hormones. *Plant Cell Rep.* 31(4): 661-669.

[44] Kulkarni PA, Prabu GR and Devarumath RM (2013) Isolation and in silico depiction of novel R2-R3 MYB transcription factors from sugarcane. *Advance Biotech* 12(11): 1-7.

[45] Kulkarni PA and Devarumath RM (2012) In silico novel MYB transcription factor [EaMYB2R] from Erianthus arundinaceus. *Online Journal of Bioinformatics* 13(1):167-183.

[46] Vaze A, Nerkar G, Pagariya M, Devarumath RM and Theertha Prasad D (2010) Isolation and PCR amplification of genomic DNA from dry leaf samples of sugarcane. *Int. J. Pharm. Bio. Sci.* 1(2): 1-6.

[47] Kalwade SB, Devarumath RM, Kawar PG and Sushir KV (2012). Genetic profiling of sugarcane genotypes using Inter Simple Sequence Repeat (ISSR) markers. *Electronic Journal of Plant Breeding* 3(1): 621-628.

[48] Devarumath RM, Kalwade SB, Kawar PG and Sushir KV (2012) Assessment of genetic diversity in sugarcane germplasm using ISSR and SSR Markers. *Sugar Tech* 14(4): 334-344.

[49] Devarumath RM, Kalwade SB, Bundock P, Eliott FG and Henry R (2013) Independent target region amplification polymorphism (TRAP) and single nucleotide polymorphism (SNP) marker utility in genetic evaluation of sugarcane genotypes. *Plant Breeding* 132(6): 736-747.

[50] Li G and Quiros C (2001) Sequence related amplification polymorphism a new marker system based on simple PCR reaction, its application to mapping and gene tagging in Brassica. *Theor. Appl. Genet.* 103: 455-461.

[51] Hu J and Vick BA (2003) Target region amplification polymorphism, a novel marker technique for plant genotyping. *Plant Mol. Biol. Rep.* 21: 289-294.

[52] Bundock PC, Eliott FG and Ablett G. (2009) Targeted single nucleotide polymorphism (SNP) discovery in a highly polyploid plant species using 454 sequencing. *Plant Biotechnol. J.* 7(4): 347-354.

[53] Bundock P, Casu R and Henry R (2012) Enrichment of genomic DNA for polymorphism detection in a non-model highly polyploid crop plant. *Plant Biotechnol. J.* 10: 657-667.

[54] Kalwade SB and Devarumath RM (2014) Functional analysis of the potential enzymes involved in sugar modulation in high and low sugarcane cultivars. *Applied Biochem. Biotechnol.* 172:1982-1998.

In: Current Status of Sugarcane Research in India
Editors: A. K. Tiwari, M. Lal and A. K. Singh
ISBN: 978-1-63463-458-8

Chapter 4

Response of Sugarcane to Abiotic Stresses and Management

R. Gomathi*, S. Vasantha, S. Venkataramana, P. N. Gururaja Rao and P. Rakkiyappan
Sugarcane Breeding Institute(ICAR), Coimbatore

Abstract

Abiotic stresses are the most important limiting factors for cane productivity. These stresses include drought, salinity, waterlogging and temperature extremes, which cause adverse effects on plant growth and yield. In India, the productivity losses due to various abiotic stresses vary from 20 to 50%. Continuous irrigation with saline water, improper drainage and practical difficulties in reclaiming saline soils lead to considerable yield losses. In Maharashtra, a high recovery zone, large areas have gone out of cultivation due to salinity, alkalinity and waterlogging. Drought coupled with water logging i.e. early drought and subsequent water logging in Bihar, U.P. and Orissa is a serious productivity constraint affecting considerable area under sugarcane cultivation. Drought and temperature stress occur alone or in combination at any stage in plant development, causing reduced the cane weight and yield loss. It is known that exposure to one kind of stress usually involves an increased tolerance to other stresses given that similar effects are shared at the cellular level. Understanding the mechanisms involved in the response of plants to adverse environmental conditions is, without a doubt, the first step in the generation of crops with higher tolerance to stress. Research at the level of genes (genomics), proteins (proteomics), metabolites (metabolomics) and individuals (physiology, systemic- biology) has been fundamental in the current understanding of the response of plants to stress. Hence, sugarcane crop response to different environmental stresses viz., drought, salinity, water logging and extreme temperature conditions on growth, morphology, physiology , metabolic and molecular aspects are discussed in this chapter. The basic management practices which help in reducing the impact of abiotic stresses and sustain the cane productivity are outlined.

* Corresponding author Email: gomathi_sbi@yahoo.co.in.

Keywords: Sugarcane abiotic stresses, drought, salinity, water logging and extreme temperature, physiology, metabolic response, stress management

1. Introduction

Sugarcane productivity is mainly dependent on growth, sucrose accumulation and yield, and the environment in which it is cultivated. Environmental stresses include drought, salinity, temperature extremes, heavy metals and radiation which cause detrimental effects on plant growth and yield. These negative factors affect the root function, growth rates, metabolism and in extreme cases lead to dehydration and death. Also, the expected rise in global temperatures indicates that there is an urgent need to understand and improve plant tolerance to these stresses. In India, the productivity losses due to various abiotic stresses vary from 20 to 50% [1]. In Maharashtra, a high recovery zone, large areas have gone out of cultivation due to salinity, alkalinity and waterlogging [2, 3].

Irrigated or dependable rainfall areas offered high yields; however, the average yields remained low in constraint environments. Drought is the primary abiotic stress causing not only differences between the mean yield and the potential yield but also causing yield instability. Drought stress associated with high day temperature causes poor growth and high tiller mortality particularly during primary growth stage which normally coincides with summer months in tropics. High temperatures have deleterious effects on plant photosynthesis, respiration and reproduction. A small increase in temperature results in conspicuous effect on growth and survival. Elevated temperatures cause rapid loss of water resulting in dehydration. In addition, drought coupled with water logging i.e. early drought and subsequent water logging in Bihar, U.P. and Orissa is becoming a serious productivity constraint affecting considerable area under sugarcane cultivation. Sugarcane is moderately tolerant to flooding and water logging. However duration of water logging and the physiological stage at which the problem occurs determines the final yield and quality. Salinity is another major constraint in sugarcane agriculture. It is primarily due to irrigation with poor quality water (mostly saline). Continuous irrigation with saline water, improper drainage and inadequate reclamation of saline soils lead to considerable yield losses. It is therefore obvious that as Boyer [4] pointed out, the crop plants attain only about 25% of their potential yield because of these detrimental effects imposed by environmental stresses.

2. Sugarcane Under Moisture Stress

Water stress remains an ever growing problem and it is the major limiting factor in crop production worldwide [5]. In India, nearly 60% of the total sugarcane agriculture suffers from lack of adequate water supply mainly because of limited availability of water for irrigation in lift irrigated areas, canal closure during summer in many of canal irrigated tracts, and drought which occur in a cyclic manner [6]. Therefore water stress of varying degrees is experienced at one stage or the other of the crop growth in all most all the sugarcane growing regions of the country.

2.1. Water Requirement and Evapotranspiration

Total water requirement of annual sugarcane crop varies from 1850 mm to 2500 mm. It is estimated that 250 tonnes of water is required for production of a tonne of sugarcane. Daily evaporation in sugarcane fields varies from 8-10 mm. Solar energy, wind velocity, temperature and humidity affect the evapotranspiration. Earlier trials on response of sugarcane to irrigation suggested that maximum tonnage was obtained at Et/Ep of 0.8. Sheath moisture and moisture content of immature nodes also served as useful indices for determining the water requirement of sugarcane crop. For high yield, sheath moisture index at 5th month stage should be high enough (83 -85%), and for higher CCS%, proper drying off with sheath moisture index of about 72% at 12th month was found to be desirable.

2.2. Critical Water Demand Period

Formative growth stage (60-150 days) has been identified as the critical water demand period and stress at this early growth phase had a direct influence on the cane yield and juice quality. Yield reduction up to 60% has been recorded in a typical drought year. Water stress especially during summer months coincides with the formative phase of the crop which affects the final yield through reduction in tiller productivity, number of millable canes, individual cane weight, and finally the cane yield and juice quality [7].

2.3. Plant Responses to Drought Stress

2.3.1. Root System

Extensive root investigations revealed that the sett roots emerge from the root band (present at nodal region of sugarcane sett), and start growing within 24 hr of planting. At the third day, some roots extend at a rate of 10 mm/day and by day 5, the elongation reaches to 20 mm/day. These thin and branched sett roots are replaced by thick; fleshier and less branched shoot roots by 90 days age. Rooting depth, distribution and activity are generally affected by soil water relationships [8]. Generally more root mass occur at less than 50 cm depth in normally irrigated condition while under stress, roots penetrate vertically downwards in the form of a rope. The root system also shows penetrating type roots which reach out for water source and hence longer and thicker roots are seen under drought [9]. The varieties selected for greater rooting depth suffered the least water deficits as compared to the normally irrigated plants. However, reports of diminished root development under moisture stress has been reported by Sheu and Kong, [10, 11, 12, 13]. Differences in root growth were related to differences in growth of susceptible and tolerant varieties [14].

2.3.2. Shoot System

The maximum LAI is generally achieved by about 6 months from planting and then slowly declines. A high LAI produces large structural apparatus for the production of photosynthate and a higher yield. Leaf expansion is very sensitive to stress. Large differences occur in the density of stomata of crop plants. The activity of stomata is greatly affected by external factors such as light, temperature, and humidity. Direct sunlight makes stomata to open, while weak and diffusive light result in closure. This explains the beneficial effect of early morning sunshine on sugarcane. Since drought is common in many sugarcane growing areas, it is important to consider reducing the transpiration and thereby reducing consumptive water use. Transpiration occurs predominantly (>90%) through the leaves while nodal region, which is free from wax deposition. Significant reduction in water loss (10 to 20%) was demonstrated due to passive curling of leaves, which reduce the radiation receipt by leaves thereby reducing water loss and increasing water use efficiency to a greater extent. Cell growth is retarded under mild stress which in turn results in reduced leaf area, followed by reduced sink growth and reduced stem elongation. The major attribute is the drying off of older leaves and stunted growth of stem resulting in a dwarf canopy. The young leaves however remain green, but when the stress intensity becomes severe, the entire crop loses its turgidity and drying will be hastened. Characters like leaf thickness, leaf dry weight and leaf area ratio are highly sensitive to drought. Deposition of wax, which is a protective mechanism, is also seen on the upper surfaces of the sugarcane leaves and stem.

2.3.3. Light Interception and Photosynthesis

Sugarcane is one of the most efficient crops capable of converting a maximum of 2-3% of solar energy into organic matter through an efficient photosynthetic system [15]. It has been estimated that one hectare of sugarcane can produce 100 tonnes of green matter which is more than twice the yield of most other commercial crops [16]. Majority of the clones intercepted 60-80% of the radiation at the completion of formative phase. The light falling on the crop surface varied from 1275 to 1950 μmol m^2 s^{-1}. In the initial stage of the stress, stomatal closure occurs which reduces transpiration rates and a decrease in leaf water potential which collectively influence the photosynthesis and productivity. The chlorophyll content also decreases resulting in low CO_2 fixation. Drought during the vegetative period tends to slow down the leaf development and canopy expansion. Chlorophyll fluorescence kinetics changed significantly during moisture stress indicating that photosynthetic electron transfer system (PETS), especially PSII and carbon assimilation were inhibited [17].The decrease in chlorophyll fluorescence was related to drought tolerance of varieties [18]. Leaf water potential and stomatal diffusive resistance are the measure of stress intensity and were found to be related to the yield of a variety. These two parameters which were identified as water stress indicators were found useful for screening varieties for drought resistance [19, 20]. The carbon isotope discrimination at 240 days was negatively and significantly associated with leaf area and total dry matter, but with photosynthesis and transpiration, the relationship was not significant [21].

2.3.4. Dry Mass Accumulation and Distribution

Sugarcane has the capability of producing 65 MT of above ground dry mass per year. The dry mass production rate ranged from 20 to 35 g / day during active growth phase and the energy conversion efficiency was estimated to reach a maximum of about 1.8% [22]. The increase in dry matter was low during periods of incomplete canopy development. The average dry matter produced was 16.83, 41.23, or 49.41 tonnes/ha or 4.81, 22.41 and 47.48 tonnes /ha under drought at the completion of formative (150days), grand growth (240days) and maturity (360 days), respectively [9].

The growth analysis studies indicated that net assimilation rate (NAR) and relative growth rate (RGR) were high during early growth phase, but declined with the age of the crop. Leaf area ratio (LAR) and leaf area index (LAI) increased with crop growth under normal irrigation while drought caused 34.62% reduction in LAI [23]. Harvest index was significantly associated with cane yield, sugar yield and CCS% [24].

2.3.5. Biochemical Responses

Sugarcane plant responds to the stresses at the biochemical level. The cellular water deficits results in the concentration of solutes, loss of turgor, change in cell volume, disruption of water potential gradients, change in membrane integrity, denaturation of proteins and several other physiological and molecular components. The concentration of malondialdehyde, a lipid peroxidation product doubled as the leaf water potential declined [25]. Epicuticular wax content was significantly high in drought resistant varieties when compared to drought susceptible types. Cellular membrane thermo stability and electrolyte leakage decreased due to water stress thereby increasing the membrane injury to as high as 85% in susceptible types. Drought tolerant varieties recovered effectively during rehydration (>60%).The capacity to maintain high membrane themostability is an important feature of tolerance to water stress [26].

2.3.5.a. Osmoprotection

Recently, interest has been generated on osmotic adjustment, turgor maintenance and growth. Turgor can be maintained by increasing various osmolytes. Accumulation of osmolytes (proline, glycine-betaine, polyamines, sugars *etc.*) which maintain the turgor and reduce the osmotic potential, help the plant to cope with the drought effect, the phenomenon called as osmoregulation.

Concomitant with 70% reduction in leaf water potential, the osmotic potential increased in many varieties suggesting an increased accumulation of osmolytes. Under water deficit conditions, the proline accumulation increased several folds in sugarcane and a significant varietal variation was noticed by Rao and Asokan [27]. Drought stress leads to the generation of reactive oxygen species (ROS) which include superoxide anion radicals (O_2), hydroxyl radicals (OH), hydrogen peroxide (H_2O_2) and singlet oxygen ($O^{\cdot}$) which cause damage to the cellular system. Drought enhanced activities of peroxidase and polyphenol oxidase have been reported in popular cultivars of sugarcane [28]. The process of osmoprotection prevents protein denaturation, helps preserve enzyme structures and protects membranes from damage by ROS.

In several studies proline accumulation was used as a screening test for drought resistance. Another metabolically inert compound called betaine also accumulates under stress. Carlin and Santos [29] evaluated the sugarcane variety IAC91-5155 under water stress and observed a trehalose accumulation of 25.9% (increase of 0.54 μmol g^{-1} fresh mass weight) at the 60th day under stress, reaching concentrations of trehalose of 2.54 μmol g^{-1} of the fresh weight. Queiroz1 et al., [30], reported the increase in levels of trehalose and free proline found to confirm what many others have reported: the importance of the osmotic adjustment of plant species, genotypes and cultivars to water deficiency in the soil.

2.3.5.b. Nutrients

Drought imposed during formative phase significantly reduced P content while N and K did not decrease [31] contrary to the earlier report of decreasing N and K content by Samuels [32].

2.3.5.c Abscisic Acid

Abscisic acid (ABA) accumulates in drought-affected leaves. ABA content enhances the leaf water potential by 1 to 2 bars and thus helps in dehydration postponement. The ABA was also found to possess a direct and stabilizing effect on protoplasm, and drought induced senescence of leaves. Dry matter production by ABA treated plants was greater than that of control. This was due to a greater development of shoot at the expense of roots. External application of abscisic acid ($1x10^{-5}$ M) exerted a regulatory role on stomatal diffusive resistance and helped in maintaining relatively high water potential [33]. ABA content enhanced the leaf water potential by 1 to 2 bars and thus helped in dehydration postponement and drought induced senescence of leaves.

2.3.5.d. Enzymes

Enzymes such as nitrate reductase (NRase), sucrose phosphate synthase (SPS), invertase etc. have been found to be regulated by the tissue water status. Nitrate reductase activity is reversible and the extent of loss under stress is to an extent of 30% and the regulation of nitrogen metabolism and the constituent end products are affected in the rate limiting way. Moisture stress induced reduction in the activity of SPS and sucrose synthase was reported in popular cultivars of sugarcane, which on rehydration resumed to normal level [34].

2.3.5.f. Sucrose Accumulation

The sucrose content in cane will be high during maturity period in the normal crop as compared to the stressed cane. Sucrose accumulation increases by about 100% while cane tonnage increases by only 20% during the maturity phase [35]. Sucrose accumulation begins at the bottom of the stem and progresses upward to the top internodes. After about 11 months age, the sucrose % in the normal crop remains constant while the percentage in the stressed crop continue to increase until 14 months.

2.3.5.e. Cane Elongation

Cane elongation is positively correlated with amount of irrigation water [36]. The rapid cane elongation (60 to 70 %) takes place during grand growth during which the seasonal available water will be utilized. Large reduction in stalk number, height, cane yield and sucrose yield were noticed due to drought. Shih and Gascho [37] reported that stalk elongation was positively and strongly correlated with water content of the elongating and meristematic tissues and cumulative soil water depletion.

3. Salinity and Sugarcane

Soil salinity threatens agricultural productivity in 77 mha of agricultural land, including 45 mha (20% of irrigated area) is irrigated and 32 mha (2.1% of dry land) is unirrigated [38].

Sugarcane is grown in India in about 5.02 million hectares, and about one fourth of the acreage is affected by salinity, alkalinity and (saline) irrigation water. The salts that largely contribute to salinity include the chlorides and sulphates of sodium, calcium, magnesium and potassium. The electrical conductivity of these soils is more than 4dS/m, while alkalinity is imparted mainly by sodium carbonate. In such soil the plants are unable to absorb the water and nutrients in adequate quantities due to high osmotic pressure of the soil water. Sugarcane is ranked moderately sensitive to salinity with a threshold value of 1.4 dS m^{-1} [39]. Soil root zone EC below 2dS,m-1 have no effect on growth and yield: 5-7.0, the yield decreases by 50 % and at EC of 8.0, stools of some cultivars are killed and do not survive. A yield reduction of up to 60% has been recorded due to salinity.

3.1. Relative Salt Tolerance of Sugarcane at Various Growth Stages

Various experiments conducted over the years showed sett germination (bud sprouting) to be the most resistant phase whereas shoot growth following germination being the most sensitive phase to salinity. The severe sensitivity of sugarcane to salinity at various growth stages is manifested by a considerable reduction in growth rate [40]. Salinity reduced tillering and other growth parameters, leaf/shoot elongation being the most sensitive and leaf/internode number being the least sensitive parameter. Sugar accumulation in the canes, even though invariably reduced might not show its effect upon juice analysis of the harvested canes in terms of sucrose % juice because reduced internode growth at moderate levels of salinity may compensate for reduced accumulation of sugars.

3.2. Sett Germination

Germination is delayed progressively with increasing salinity and reduction in final germination percent observed at higher salinity levels (EC >5dSm^{-1}). Growth of leaf blades showed a maximum reduction compared to stem and sheath whereas that of sett roots was the least affected during sett germination phase [41]. Reduced germination with biomass variation for root and shoot has been well documented in several works [42,43,44]. Soil salinity has a profound impact on the crop growth specially so with the process of germination. Germination was delayed under salt treatment and reduction in final germination percent was observed at higher salinity level (EC >5dS,m^{-1}). Higher reduction in germination of setts with increasing salinity levels were reported for sugarcane. Kumar and Naidu, [44] observed that soil salinity as more damaging for germination of setts at low temperature (below 25° C). Varietal response is a critical factor in determining the final germinant. For instance genotypes like Co 97010, Co 95007 etc. recorded a reduction in germination over 50% indicating their sensitiveness.

3.3. Tillering and Early Growth

Tiller production per main shoot decreases under saline as well as sodic conditions. In a study with 10 popular varieties, the reduction in tiller production due to salt treatment was from as low as 16.3% in Co 6304 to as much as 49.8% in Co 86010. Consequently, shoot population was also reduced resulting in poor and patchy field stand. Shoot height, number of internodes, number of leaves and leaf area per plant were significantly lesser in saline soil. Decreased or nil expansion growth of leaves and young internodes results in stunted canopy and poor tillering results in poor crop. Apart from tillering, cane formation was inhibited and the internodes were very narrow suggesting the sensitivity of expansion growth. The reduction in tillering due to soil salinity as well as high salt concentration in irrigation water has been reported in sugarcane [45]. Restricted growth in terms of reduced shoot height and less green leaf production for photosynthesis was reported [46, 47, 48, 41]. Shoot growth rate reduced even under mild salinity (EC of $2dSm^{-1}$) in different cultivars of sugarcane [49]. Ion-toxicity was the main determinant of salt tolerance at the grand growth stage while the osmotic component of NaCl mainly appeared to affect the transport of sucrose to stalks, followed by stimulated sucrolytic activity in the internodes, resulting in reduced final cane yield [50].

3.4. Yield and Quality Characters as Influenced by Salinity

The cane maturity is delayed by salinity. In some genotypes the juice quality is severely affected so also the sugar yields [51]. Reduction in number of millable canes was up to 37% in popular genotypes with tolerant genotypes recording lesser reduction. Cane length, girth, number of internodes showed reduction due to salt treatment, which ultimately reduced the cane weight and yield by 38.56 percent Gomathi and Thandapani [52]. However the extent of reduction was found to be less in tolerant varieties viz., C 92038 and Co 85004 (29.81 and 28.00 %) compared to susceptible varieties viz., Co 85036 and Si 94050 (47.82 and 47.36 %).Cane yield recorded significant reduction of up to 64% in sensitive genotypes while it was marginal (27%) in tolerant types. A decrease in cane yields of the order of 5.45 t/ha for every 1 dSm^{1}/ha is experienced due to soil salinity [53].

3.5. Juice and Jaggery Quality Characters as Affected by Salinity

Sucrose% juice, brix and purity are reduced by salinity. Increased non-sugar solids and salts reduce the purity. The salt content of cane juice ranges from 900-1900 ppm in non-saline soils whereas it ranges from 4000-4500 ppm in saline soils. The electrical conductivity of the juice at harvest increased in all the genotypes under saline conditions due to irrigation with saline water; increased accumulation of Na, K and Cl ions caused a reduction in sucrose per cent juice due to salinity. Concentration of Na is generally below 10 mM whereas that of K and Cl may go up to a maximum of 150 mM under saline conditions (EC 7.5 dSm-1). K and Cl concentrations were negatively correlated with sucrose % and purity % of the juice and stalk diameter but were positively correlated with number of millable stalks in inter specific hybrids (ISH) clones tested. Major and micro nutrient uptake and partitioning of the essential nutrients viz., N, P, K, Ca, Mg, Zn, Fe and Mn were estimated in plants exposed under long term salinity stress [54], Results showed that the tolerant genotypes (C 92038 and Co 85004) maintained higher N, P, K, Ca Mg, and higher K/Na ratio compared to susceptible genotype

(Co 85036). Accumulation of toxic elements was noticed in susceptible genotypes viz., Na, Cl, Bo, Mo, which resulted in expression deficiency symptom of Ca.

The quality of the jaggery is dependent on the cane juice which in turn is determined by the variety and the environment in which the cane is grown. Well pronounced differences in jaggery quality as indicated by net rendement value and colour were observed among the tolerant genotypes. Under high soil salinity, the tolerant genotypes Co 85019, Co 94008 and Co 97008 produced jaggery with poor grade, colour and taste while, the genotypes Co 94012 and Co 99004 produced good quality jaggery even under salinity as sodium and chloride content increased only marginally and cane yield and juice quality were not affected. In the context of sizeable area of sugarcane being grown under saline soils, there is a need for identification of genotypes like Co 94012 and Co 99004 able to produce good quality jaggery under saline conditions. Content of Na in juice is to be considered essential new criteria than salinity tolerance per se in identifying genotypes' suitability exclusively for jaggery making purpose [55].

3.6. Physiological and Metabolic Behaviour under Salinity

Salinization leads to a decrease in rates of transpiration, stomatal conductance and CO_2-assimilation of all the leaves present on the plant. The damaging effect increases with time after salinization. Gas-exchange measurements suggested that variation in carbon isotope discrimination (delta) was attributable largely to variation in bundle sheath leakiness to CO_2. Salinity-induced increases in (phi) appeared to be caused by a reduction in C3 pathway activity relative to C4 pathway activity rather than by physical changes in the permeability of the bundle sheath to CO_2 [49]. The rates of transpiration continue to decrease probably due to its effect on stomatal conductance whereas it was not the case with rates of assimilation. The effect appears to be due to their effects on its efficiency to fix CO_2 present in the leaf rather than its deficiency.

Accumulation of sugars in the leaves upon salinization appeared to result from reduced rates of their translocation, which in turn appeared to be related with their reduced utilization in the sink tissues. Thus reduced rates of photosynthesis were not directly responsible for reduced growth under saline conditions. Results of another experiment at grand growth phase, the tolerant clones maintained more or less uniform rates of photosynthesis, while the sensitive types showed sharp decline due to salt treatment. Net photosynthetic rates were reduced when the leaf water potential was <-0.9 MPa on diurnal basis, suggesting the sensitivity of the photosynthesis process to water potential gradients. A reduction in stomatal conductance may result from the osmotic effects of salinity. Net photosynthetic rate reduced when the leaf water potential was <-0.9 MPa on diurnal basis, suggesting the sensitivity of the photosynthesis process to water potential gradients. Fluctuations in photosynthetic rate during hours of day in stress free environment and stressful environment would account for the variation in net photosynthetic rate and photosynthate production.

Long term salinity effects were studied in tolerant and sensitive genotypes of sugarcane in order to understand the mechanism of salinity tolerance. Water and osmotic potentials were distinctly different between tolerant and sensitive genotypes during grand growth phase (150-240 DAP). The tolerant sugarcane genotypes viz., C 92038 and Co 85004 showed higher osmotic adjustment by lowering of Ψ_l and π_l due to more accumulation of free proline, polyols and TFAA resulting in more uptake of water from external solution, and thus TP and RWC were maintained under salinity condition [56]. Result of correlation study were also indicated that the P was positively associated with RWC, rs, FP, polyol and TFAA and it negatively associated with $\Psi_{l,}$ π_l and Tr under long term salinity, which showed the role of osmolytes in degree of osmotic adjustment under salt stress condition.

Reports are available that, long term maintenance of water status, osmotic adjustment, maintenance of high photosynthetic rate and biomass production are essential features for a sugarcane genotype to perform as tolerant type [56, 57]. Total biomass on an average was reduced by 41% under salt treatment with tolerant clones showing only a moderate reduction of 28 and 17% during formative and grand growth phases respectively, whereas, the sensitive clones showed reductions of 60 and 71% at formative and grand growth phases respectively [57].

3.7. Lipid Peroxidation and Cell Membrane Injury

Malondialdehyde (MDA), a lipid per oxidation product, varied from 0.85 $\mu g\ g^{-1}$ to 1.667 $\mu g\ g^{-1}$in control while it varied from 1.28 to 2.51$\mu g\ g^{-1}$ under saline conditions. Tolerant genotypes recorded lesser average increase of ~28% in MDA while sensitive genotypes recorded nearly double the increase of ~57% thereby indicating for a greater damage to the membrane system. Cell membrane injury test conducted with popular varieties showed significant variation in their tolerance capacity. Cell membrane stability is a measure to test the membranes biophysical /biochemical properties. Under stress situations, the cell membrane loses the selectivity of ions and macromolecules resulting in heavy influx/efflux of essential ions from the cells. Cell membrane injury test conducted with popular varieties showed significant variation indicating their tolerance capacity. In tolerant genotypes the MDA content increased by 36% while in sensitive genotypes the increase was 57% [58].

3.8. Oxidative Enzymes under Salinity Stress

The role of plant oxidants systems in salt stress tolerance was studied in four contrasting sugarcane genotypes. Salt stress imposed at different stages of crop growth resulted in an increase in lipid peroxidation and decrease in membrane stability, chlorophyll florescence ratio (fm/fv) and chlorophyll and carotenoide contents. The antioxidant enzymes ascorbate peroixdase, glutathione reductase, and superoxide dismutase also increased significantly under salt stress. It seems that salt tolerance of C 92038 and Co 85004, as represented by higher membrane stability and chlorophyll and carotenoide contents, fm/fv ratio and lower lipid peroxidation, is related to its higher antioxidant enzyme activity [59].

Two low molecular forms with faster mobility were induced under higher salinity level only in tolerant genotypes, suggestive of its role in tolerance behavior and isolated chloroplast lysate also showed induced isoforms under high salt condition [54, 58, 59]. SOD activity increased marginally in response to high salt condition in varieties Co 85019 and Co 95003 and in other varieties activity was on par with control plants. SOD isoforms (five in all) were similar in both control and salt treatment. Either treatment or genotypic influence could not be detected in isoforms of SOD or in its activity [54, 58, 59]. Reports are available that the activity of ascorbate peroxidase (APX) significantly increased by two fold in tolerant varieties while in sensitive types the increase was only marginal. However, APX isoforms failed to show any variation due to high salt treatment [54, 58, 59].

3.9. Enzymes of Sucrose Metabolism

The enzymes of sucrose metabolism viz., sucrose synthase (SS), sucrose phosphate synthetase (SPS) activity declined due to salinity. The tolerant genotypes showed relatively less reduction [52]. Progressive stress responses enlighten us about the metabolic changes during stress adaptation in tolerant types and any flaw that reflect on the metabolic failures resulting in sensitive behaviour. The salinity (NaCl) effect was noticed in the sensitive variety Co 95007, on day two with poor growth. The visual symptoms i.e., yellowing of leaves and salt injury in leaves were noticed on day seven in the sensitive variety. Progressive stress responses were studied in contrasting sugarcane genotypes to elucidate the stress adaptative features with regard to physiological and biochemical characters. Varieties showed differences with respect to parameters studied from the day four. The tolerant variety Co 85019 maintained stability of plastid pigments (chlorophyll and carotenoids), higher proline level and increased activity of oxidative enzymes viz., POX, SOD). Sensitive genotype suffered heavy loss with regard to these characters. Lipid peroxidation, a measure of damage to the membrane system was high in sensitive variety and difference between genotypes became significant from day four, indicating the progressive nature of adaptation in tolerant and its failure in sensitive variety [54, 59, 60].

3.10. Salt Induced Proteins and Nucleic Acid

The proteins which accumulate in response to salt stress are referred to as salt induced protein (SIP), which may have some relevance in stress tolerance. The criterion for a stress protein to be considered as molecular marker is that it should show differential accumulation pattern with respect to tolerant and sensitive cultivars. Under salt stress, specific expression of salt shock protein with molecular weight of 15 kDa, 28 kDa and 72 kDa were noted in tolerant genotypes(C 92038 & Co 85004), while it was completely absent in susceptible (Co 85036 & Si 94050) [61]. Further, they concluded that the variation in RNA content among the genotypes might be responsible for specific synthesis salt shock proteins (15 kDa, 28 kDa and 72 kDa) and it considerable adaptive significance under salinity conditions.

4. Sugarcane under Waterlogging

Water logging drastically reduces the growth and survival of sugarcane worldwide and cane yield reduction is estimated between 15-45%. A considerable area under sugarcane crop in several parts of India (Assam, Bihar, and West Bengal, and eastern UP, coastal region of Andhra Pradesh, Tamil Nadu, Kerala and Karnataka) are exposed to stagnant water for two to three months during monsoon season. It was observed that sugarcane crop was susceptible to water logging in the first 3-4 months, somewhat tolerant at 4-9 months age and helped by it in maturity beyond that age. Higher water table during active growth phase adversely affects stalk weight and plant population resulting yield loss at the rate of about one ton per acre for one inch increase in excess water [62, 63].

Some physiological effects of cane are found due to water-logging are i) transpiration rates are reduced due to stomatal closer, ii) rate of photosynthesis is considerably reduced presumably that cause the reduction of effective leaf area, iii) growth rates are drastically reduced during water-logging, iv) higher respiration rate of submerged organs compared to leaves. A shift in respiratory metabolism from aerobic to anaerobic pathways is one of the main effects of oxygen deficiency causing from waterlogging. The effects of water logging on respiration rate depend on the varieties and its physiological age. It is also reported that under waterlogging condition some morphological, anatomical, physiological and biochemical changes take place in plant for sack of adaptation/ survival [64, 65].

4.1. Germination

Studies conducted in Australia indicated that waterlogging decreased germination if soil was saturated for more than 3 days [66]. Flooding at planting affects emergence. A sugarcane variety Cp 89-2376, less than 6 days of flooding has more emergences and CP 72-2086 had externally low emergence [67].

4.2. Root System

In the absence of oxygen, root hairs die and eventually the roots blacken and rot with the results entire underground root system gets choked and root respiration is also impaired. Because of the insufficient and inadequate root system absorption of nutrients and water is seriously affected. Nutrient absorption is further affected by their unavailability. During natural water logging of the soil, roots exposed to hypoxia condition under such situation roots able survive either by inducing biochemical acclimation or by anatomical acclimation. Following sensing of partial oxygen deficiency, genes coding for so called anaerobic proteins (HIPs protein) are up –regulated at transcriptional and post- transcriptional levels and the HIPs are necessary for the acclimation [68].

4.3.a. Anatomical Acclimation through Aerenchyma Formation

Sugarcane is supported by adventitious roots in waterlogging conditions, which develop possibly as result of hormonal imbalance induced by hypoxia and decreased supply of oxygen to the submerged tissue. These roots remain in the upper layers of the water which are presumably richer in oxygen content. These are adapted to water logging conditions than the original roots because they have much larger intercellular spaces [69]. Study examining genetic correlation of sugarcane traits under flood, found that selection for adventitious root development may not increase sugarcane yield under flooding [70]. All these supportive roots developed during flooded situations, helped in maintenance of root activity by supplying necessary oxygen [71].

4.3.b. Role of Ethylene

Ethylene involvement in adventitious root formation and aerenchyma formation in several crops was summarized by Jackson and Richard [68]. Ethylene synthesis increases under flooded conditions when ACC synthase concentrations increase, which stimulates ACC synthesis. ACC then diffuses to aerated parts of the root and is converted into ethylene by ACC oxidase. Ethylene is far less soluble in water than in air. Therefore, more ethylene is retained inside plant tissues when flooding occurs and ethylene concentrations increase.

4.4. Shoot Growth

It is inevitable that, because of the close functional inter dependence between roots and shoots, stress on roots from water logging also threatens the shoot system. One example of this is the arrest of nitrate uptake that arises from microbial de nitrification and damage to uptake mechanism from an absence of oxygen. Young leaves remobilize the nutrients from older leaves leading to premature senescence of the later [72]. The impact of water logging on shoot growth can be observed on changes in growth habit, visual health, internal anatomy, water relations, hormonal and nutritional composition. Water logging can inhibit leaf and stem expansion and tiller production and cause epinastic curvature of petioles, orientation of shoot extension [73]. Sarkar et al., [74] was observed that plant height after 12 days of submergence showed significant positive association with survival percentage. In sugarcane, varieties which maintained better shoot height and internodal length yielded better under flooding condition [75, 65].

4.4.a. Tiller Production

Flooding during tillering, resulted in greater tiller mortality and reduced stalk population. Flooding at any stage reduced production of new tillers and rate of elongation of the established tillers [76] and decrease was more with longer duration of flooding. Varieties also differed in the regard [77]. Studies conducted at SBI, Coimbatore, indicated that the waterlogging stress during formative phase of the (90-170 DAP) caused 13.00, 21.63 and 26.52 % reductions in plant height, tiller production and leaf area respectively. However, the reduction was less in the resistant clones [78].

4.4.b. Leaf Development and Growth Parameters

The expression of yellowing symptom in leaf, higher stalk mortality, faster drying of lower leaves, reduction in leaf number and size, are the morphological changes due to waterlogging stress. Further, waterlogging resulted in 26.5%, 25.2% and 24.0% mean reductions in , leaf area index (LAI), leaf nitrogen content and total chlorophyll content, respectively [65, 56, 79].

4.5. Yield and Quality

Cane yield losses depend upon the duration of water logging, stage of crop growth and management practices before, during and after water logging. Yield loss occurs due to stalk mortality, reduced crop growth due to lack of nutrition and water uptake, lodging, cane breakage, etc. About 5-30 % loss in yield was reported for 15-60 days of water logging condition created artificially during the late grand growth phase (7.5-9.50 months). A study conducted in tropical India (Tamil Nadu) indicated that in a water stagnation period of 2 months the reduction in cane yield was to the tune of 26-36% in various varieties [80]. A study conducted at SBI, Coimbatore showed that the waterlogging caused 22.4%, reduction in NMC, 45.6 % reduction in single cane weight, 30.0% reduction in cane length, 15.9% reduction in internodal length, 17.8% reduction in cane thickness and 40.1% reduction in cane yield [56].

4.6. Photosynthesis and Partitioning of Assimilates

Under anaerobic condition, photosynthesis declined due to slow diffusion of CO_2 in water and reduced availability of light as result flow rate of assimilates to the roots also decreased. In sugarcane, chlorophyll content reduced under submergence and the reduction was more pronounced in susceptible varieties results in reduction photosynthetic rate and leaf dry matter accumulation [56]. Reports available in sugarcane that the reduced photosynthesis after lowering of the water table to end saturation caused a reduction in biomass by harvest date in the range 40 to 50% in saturated treatments as compared with the control [81, 82].

4.7. Anaerobic Proteins (ANP'S) in Response to Flooding Stress

Plants also respond to anoxia by altering the pattern of root protein synthesis. The proteins which are synthesized as a specific response to anaerobiosis are called the anaerobic polypeptides (ANPs) [83]. Flooding stimulated the synthesis of a small group of proteins known as anaerobic polypetides (ANP's) appear to play an essential role for anoxia survival. All the characterized polypeptides are glycolytic enzymes [84]. Among the ANPs, ADH is predominating one and has been extensively studied [83]. New synthesized ADH isozymes emerge during flooding in many plants [85, 86] and with different biochemical properties. Both in leaf and root, specific expression of ANP's viz., 66 kDa, 98 kDa and 132 kDa proteins in response to short term flooding stress was recently reported in sugarcane especially in tolerant genotypes (Co 99006 and Co 8371), indicating their possible role in tolerant behaviour [56].

4.8. Nitrate Reductase Activity

Reduction of NRase in leaves of waterlogged plants results in the rapid depletion of the nitrate and oxygen is consumed by soil biota and then anaerobic conditions develop. Gomathi and Chandran [65] found a positive association between Nitrogen content of index leaf and nitrate reductase activity (NR ase) with flooding tolerance of sugarcane clones exposed to long term flooding stress.

4.9. Alcohol Dehydrogenase (ADH)

Alcohol Dehydrogenase (ADH) is responsible for the synthesis of alcohol and regeneration of NAD in alcoholic fermentation [87, 88]. This regenerated NAD enables glycolysis to continue under anoxia, thus producing a net 2 moles of ATP per mole of glucose relative to the 38 moles of ATP produced under aerobic conditions through respiration. A significant increase in ADH activity was recently reported in sugarcane due to short term flooding [56].

4.10. Antioxidant System in Response to Waterlogging

Tolerance to wide varieties of environmental stress conditions has been correlated with increased activity of antioxidant enzymes and levels of antioxidant metabolites (Davies 1987). A short-term waterlogging treatment led to an increase in the activities of anti-oxidant enzymes viz., APX, CAT and SOD in sugarcane [89]. Weijun Zhou and Xianqing Lin [90] concluded that Leaf chlorophyll content and SOD and CAT activities were markedly reduced after plants were waterlogged for 30 days at various stages of growth and results indicated that waterlogging could promote the degradation of chlorophyll, reduce the activities of SOD and CAT, and therefore accelerate leaf senescence.

5. SUGARCANE RESPONSE TO TEMPERATURE EXTREMES

5.1. High Temperature

Temperature is a major environmental attribute influencing the crop yields. While optimum temperature is necessary for growth, metabolism and final productivity, high temperature induces a wide variety of changes in cellular structures and metabolic process. When the magnitude and duration of the heat stress exceeds a threshold, cells are irreversibly damaged and die. Different living systems respond differently to increased temperature. In addition to speeding up of phonological events, high temperatures have deleterious effects on photosynthesis, respiration and reproduction including seedling survival. A small increase in temperature can have pronounced effect on growth and survival. A 10° rise in temperature may be sufficient to kill the plant. Temperature stress in plants is dependent on many factors including the thermal adaptation of the species or genotypes, the duration and the growth stage of the exposed tissue.

High temperatures caused significant declines in shoot dry mass, relative growth rate and net assimilation rate in maize, pearl millet and sugar- cane, though leaf expansion was minimally affected [91, 92] Major impact of high temperatures on shoot growth is a severe reduction in the first internode length resulting in premature death of plants [93]. For example, sugarcane plants grown under high temperatures exhibited smaller internodes, increased tillering, early senescence, and reduced total biomass [94].

5.1.1. Heat Resistance

Plants normally try to avoid high temperature by adjusting the canopy architecture through insulation, lowered respiratory rates, decreased absorption of radiant energy transpirational cooling etc. Avoidance or tolerance of direct heat injury can be achieved by different means such as presence of substances that inhibit coagulation, and increased alkalinity of the protoplasm, the decrease in free water, greater thermostability of proteins, enhanced protein bond strength, increased concentration of sugars etc.

5.1.2. Heat Stress Acclimation and Adaptation

Plants manifest different mechanisms for surviving under elevated temperatures, including long-term evolutionary phenological and morphological adaptations and short-term avoidance or acclimation mechanisms such as changing leaf orientation, transpirational cooling, or alteration of membrane lipid compositions. In many crop plants, early maturation is closely correlated with smaller yield losses under high temperatures, which may be attributed to the engagement of an escape mechanism [95]. Inadequate responses at one or more steps in the signaling and gene activation processes might ultimately result in irreversible damages in cellular homeostasis and destruction of functional and structural proteins and membranes, leading to cell death [96,97].

Adaptive response of settling and calli by heat acclimation were studied in Co 86032, Co 99004, Co 2001-13, Co 2001-15, CoM 265, Co 218 and Co 315 by heat acclimation both under *invivo and invitro* conditions [89]. Results showed that the in all the varieties, callus and settlings that were induced for high temperature recorded higher soluble protein, proline, glycine betaine, total phenols content and ROS enzyme activity compared to non induced callus and settlings. Exposure to high temperature caused a significant increase in lipid peroxidation (MDA content) and cell membrane injury (%), however pre treated settlings and calli recorded lesser cell membrane damage. Different fractions of heat shock proteins (*hsps*) were expressed upon heat acclimation (90 kDa, 70 kDa & 27 kDa). Varieties, Co 218, Co 315 and Co 99004 performed better in terms of accumulation of metabolites, ROS synthesizing enzymes and expression of specific proteins for temperature induction response (TIR) compared to CoM 265, Co 2001-13 and Co 2001-15 [98].

Heat acclimation pretreatment improved the thermotolerance of settlings and calli of Co 86032 compared with those without heat acclimation pretreatment under heat stress, which may result from decrease in membrane lipid peroxidation and accumulation of ROS, an increase in activities of antioxidant enzymes (APX, POX, SOD), accumulation of metabolites (total phenolics, soluble proteins & sugars, proline, glycine-betaine etc.) and specific expression of heat inducible proteins (Hsp90, Hsp70) and dehydrins (27 kDa) [61]. Therefore, these physiological changes caused by heat acclimation may be helpful to improve the heat stress adaptation of sugarcane settlings and calli. One of the most closely studied mechanisms of thermotolerance is the induction of hsps, which, as described in above, comprise several evolutionarily conserved protein families. However, each major HSP family has a unique mechanism of action with chaperonic activity. The protective effects of hsps can be attributed to the network of the chaperone machinery, in which many chaperones act in concert. An increasing number of studies suggest that the hsps/chaperones interact with other stress-response mechanisms [99].

5.1.3. Photosynthesis and Water Relations

Several experiments have shown that all genotypes are sensitive to temperature at one stage or another and phenological stages differ in sensitivity to temperature. Among all plant processes, photosynthesis is essentially an important event ion crop growth. Temperature is influence yield by regulating the rate of biomass accumulation through photosynthesis and the duration of growth [100, 101]. The optimum temperature fo rphotosyntheis is higher in C4 plants as compared to C3 plants [102,103]. Genotypes most tolerant to high temperatures had the most stable leaf photosynthetic rates across temperature regimes or they had the longest duration of leaf photosynthetic activity after anthesis and high grain weights. Despite observed negative effects of high temperature on leaf photosynthesis, the temperature optimum for net photosynthesis is likely to increase with elevated levels of atmospheric carbon dioxide. In sugarcane an increased chlorophyll *a*:*b* ratio and a decreased chlorophyll: carotenoids ratio were observed in the tolerant genotypes under high temperatures, indicating that these changes were related to thermotolerance [104]. Under field conditions, high temperature stress is frequently associated with reduced water availability [105]. In sugarcane, leaf water potential and its components were changed upon exposure to heat stress even though the soil water supply and relative humidity conditions were optimal, implying an effect of heat stress on root hydraulic conductance [106].

5.1.4. Enzymes

High temperature causes denaturation of the protein, coinciding with drop of the function of the enzymes due to the loss of tertiary structure of the protein at high temperature. As a consequence of breaking up the weak molecular bonds that keep polypeptide chain folded appreciably. Those proteins that have disulfide bonds in tertiary structure are relatively resistant to denaturation. Thermal stability of the enzymes can vary to some degree amongst species that grow on different environments. Differential thermal stability of isoenzymes has been reported as one of the parameters associated with high temperature tolerance. Hydroxypyruvate reductase and glutathione reductase are two thermostable enzymes, which can play important role in protecting plants from heat stress.

5.1.5. Osmoprotectants and Metabolites

Desiccation and dehydration due to heat stress inflict injury to the plants through destabilization of the plasma membrane. Natural osmolytes, especially soluble sugars are widely used as stabilizers of plant oligomeric proteins, protecting them against a variety of adverse conditions. Soluble sugars cause acclimation of photosynthesis to high temperature in desert plants. Higher concentration of trehalose in desiccation tolerant resurrection plants has suggested that this molecule may be involved in the desiccation tolerance of plants. Proline accumulation in plant tissues under dehydration and desiccation has indicated that it is the most potent osmoprotectant that plays a role in contracting the effect of osmotic and heat stresses. It is suggested that genetically engineered crop plants that overproduce proline under temperature stress might thus, acquire osmo tolerance that is the ability to tolerate environmental stresses such as drought and heat stress. In assessing the functional significance of accumulation of compatible solutes, it is suggested that proline or GB synthesis may buffer cellular redox potential under heat and other environmental stresses [107]. Similarly, accumulation of soluble sugars under heat stress has been reported in sugarcane, which entails great implications for heat tolerance [107, 61]. Phenolics, including flavonoids, anthocyanins, lignins, etc., are the most important class of secondary metabolites in plants and play a variety of roles including tolerance to abiotic stresses [104, 92, 107].

5.1.6. Heat Shock Proteins (HSPs)

Synthesis and accumulation of proteins designated as 'Heat Shock Proteins' (hsps) were identified due to heat stress. Subsequently it was shown that increased production of these proteins also occurs when plants experience a gradual increase in temperature more typical of that experienced in a natural environment. Three classes of proteins were distinguished based upon the molecular weight of most HSPs, namely HSP90, HSP70, and low molecular weight proteins of 15 to 30 kDa (LMW HSP). The proportions of the three classes differ among species. In general, heat shock proteins are induced by heat stress at any stage of development. Under maximum heat stress conditions, HSP70 and HSP90 mRNAs can increase ten-fold and LMW HSP increase as much as 200-fold. Three other proteins, though less important, are also considered to be heat shock proteins viz. 110 kDa polypeptides, ubiquitin, and GroEL proteins. In arid and semi-arid regions, dry land crops may synthesize and accumulate substantial levels of heat shock proteins in response to elevated leaf temperatures. The induction temperature for synthesis and accumulation of heat shock proteins in laboratory-grown cotton ranged from 38 to 41°C. Soil water deficits resulting in midday canopy temperature of 40°C or greater were used to study heat shock proteins in field-grown cotton. The mechanism by which heat shock proteins contribute to heat tolerance is still not certain. One hypothesis is that HSP70 participates in ATP-dependent protein unfolding or assembly/disassembly reactions and those they prevent protein denaturation during stress. If this mechanism is true, then heat shock proteins may provide a significant basis for increasing heat tolerance of crop plants in a global warming situation. The HSPs may play a structural role in maintaining cell membrane integrity during stress. Other heat shock proteins have been associated with particular organelles such as chloroplasts, ribosomes and mitochondria. HSPs thus provide a significant opportunity to increase heat tolerance of crops.

5.1.7. Long-Term Effects of High Temperatures on Crops

More important than acute effects of extreme temperature stress are the chronic effects of continuously warmer temperatures on crop growth and development. Record crop yields clearly reflect the importance of season-long effects on crop yields: crops generally yield the most where temperatures are cool during growth of the harvested component. This chronic effect of high temperature differs significantly from the acute effect of short-term temperature events, because seasonal temperature.

5.1.8. Screening for Thermo Tolerance

Sugarcane varietal evolution requires yield stability even under harsh climates, identification of suitable screening techniques and understanding of the physiological, biochemical processes to high temperatures is absolutely necessary. Temperature induction response (TIR), a high throughput approach has been utilized for screening of thermo tolerance in many of the crops viz., pea, pearl millet, cotton, finger millet, sunflower, capsicum, ground nut and soybean. Temperature conditions of 42 ºC with 10 h of stress treatment and 48 ºC with 20h of stress treatment were identified as sub- lethal and lethal temperature conditions for screening thermotolerant calli (*in vitro* condition) and 40 ºC with 10h and 48 ºC with 15h were identified as sub- lethal and lethal temperature conditions for screening thermotolerance in settlings (*in vivo* condition) [98,89]. The results showed that *in vivo* method was more effective in terms of time and cost than the *in vitro* method of screening for thermotolerance. Adaptive response of Co 86032 by heat acclimation was investigated under *in-vivo* and *in-vitro* conditions through TIR technique. It was found that induced settlings and calli for thermotolerance recorded higher soluble protein, proline, glycine betaine, total phenols, POX activity and SOD activity than non-induced and the above parameters indicated stability under stress treatment. The temperature tolerance is a polygenic character and many genes have been characterized, imparting respective high temperature tolerance in transgenic plants. Strategies are to be evolved for breeding high temperature tolerant crops either by hybridization or genetic engineering techniques.

5.2. Effects of Low Temperature

In most cases, plants do not suffer chilling injury until temperature drops below 10 ºC. Plants can be characterized into three categories with respect to their responses to low temperature i.e. chilling sensitive, freezing sensitive and freezing tolerant. Freezing sensitive plants are damaged by exposure to temperatures below 0 ºC. Cold and freezing temperatures have different effects on sugarcane seed pieces, young cane, and mature cane. Cold stressing (but not freezing) seed pieces for three weeks will result in reduced stalk numbers, stalk height, stalk weight and sugar yield. The results of stressing the seed cane will carry over to ratooning. Studies have shown young cane can withstand a few repeated minimal (30 °F) 2 to 4 hour freezes without resulting in reductions in stalk count and vigor. However, repeated exposure to 25 °F for 4 hours reduced stalk counts and vigor. Increasing the number of these freeze events shows additive harmful effects by the reduction of both cane and sugar yields.

Poor sprouting of stubble buds at low temperatures is associated with a lower level of reducing sugars, reduced activity of acid invertase and higher accumulation of IAA and total phenols [108]. Studies on cold tolerance were reported as early as 1960's [109], since then many studies have shown that temperature affects the growth and development of the plant in various ways. Being a tropical/subtropical crop sugarcane responds to chilling temperatures with significant alterations in photosynthesis. Severe reduction in photosynthesis has been reported [110], mainly due to reduced activities of photosynthetic enzymes and sucrose phosphate synthase [111], For young plant cane and young ratoon cane, most or all of the aboveground primary and secondary shoots that have growing points above ground will be killed by a freeze event. Post-freeze regrowth will come from secondary shoots whose growing points were below ground and protected from the cold. Early freezes on mature sugarcane slow or stop the production and storage of sucrose by the plant which may lower sugar recovery at harvest. A severe freeze causes changes in crusher juice quality which can be monitored by measuring crusher juice brix, polarity, pH and titratable acidity. With this data, sucrose concentration, purity, and theoretical sugar yield can be calculated. After freeze damage, crusher juice sucrose, purity, and sugar yield decline. At the same time, titratable acidity and gums (e.g., dextran) may increase several fold. These increases are highly variable depending on variety, maturity and growing conditions at the time of and after the freeze. Freezes can also cause a loss of sugarcane stalk weight, with time, after the freeze. Freezing temperatures cause leaf damage first. A very light freeze, between 32 and 29° F for a few hours, may only cause a banded chlorosis or burning of the leaf tips.

Changes in juice quality may become apparent in a week or two and will be variety dependent. A more severe freeze, 23 to 24° F for a few hours, will completely brown leaves and partially or entirely freeze stalks. The terminal bud and all or nearly all the lateral buds will be killed. Deterioration of juice quality may start within a few days. If the low temperature duration is short, some of the heartier varieties may not experience juice quality deterioration for two to three weeks. A freeze below 22° F will almost invariably kill all leaves, buds, and internal stalk tissue to ground level. Stalk splits will be seen, although some may be hard to detect because of closure after the freeze. Juice quality deterioration may become evident a few days after the freeze. Besides the severity of the freeze, the duration of the freeze is the next most important variable. A steady 30° F freeze for 48 hours could presumably do as much damage as a short 22° F freeze. The importance of early and continuous field damage surveys and juice quality monitoring post-freeze cannot be overemphasized. Completely frozen cane may become unacceptable to the mill within one to two weeks. Warm and humid weather following a freeze increases the rate of deterioration. Stalk damage can be determined by splitting the stalk and looking for frozen or thawed water-soaked tissue. Although standing cane routinely freezes from the top down, temperatures are normally lower near ground level and, therefore, lodged cane is more prone to damage. Thus it is evident that temperature is one of the most important abiotic factors controlling the growth, development and acclimation of natural and cultivated plant species. Plants have evolved various morphological, anatomical, biochemical and physiological means to cope with the undesirable temperature regimes. Crop cultivation is limited to temperatures operating in a narrow range of 10-30 °C. Strategies have to be evolved for breeding low temperature and high temperature tolerant crops either by hybridization or genetic engineering techniques so as to evolve the crops possessing tolerance to high and low temperature limits which are suitable for cultivation beyond traditional agriculture.

6. Molecular Interventions for Mitigation of Abiotic Stresses

A number of genes have been reported to be induced by drought, high salinity and low temperature stresses and their products are thought to function in stress tolerance and response [112, 113, 114]. The direct introduction of small number of genes by genetic engineering seems to be a more attractive and rapid approaches for improving stress tolerance [115]. Present day biotechnological strategies rely on the transfer of one or several genes that encodes either biochemical pathways or endpoints of signaling pathways. These genes/products protect either directly or indirectly against environmental stresses.

6.1. Drought Stress

Studies were conducted to detect genes expressed under drought by macroarray, identified 165 genes responded to drought. Important stress related pathways were repressed in sensitive cultivar. A great number of genes with unknown function were reported which may provide new insight into the tolerance mechanism [116]. Up regulation of genes encoding for polyamine oxidase, cytochrome-c-oxidase, s-adenosyl methionine (SAM) decarboxylase and thioredoxins, which directly or indirectly involved in the regulation of redox status was reported in sugarcane under drought stress [117]. Rodrigues et al. [116] demonstrated increased expression of a gene encoding a peroxidase in a drought tolerant sugarcane cultivar. This enzyme is responsible for the reduction of H_2O_2 to H_2O and O_2, and a decline in peroxidase activity is considered a limiting step to ROS neutralization in sugarcane [118]. The accumulation of the osmolytes trehalose and proline also contributes to the reduction in the damage caused by the accumulation of ROS and is associated with drought tolerance in sugarcane [119, 120, 121].

Area that deserves attention is the response mediated by the phytohormone absisic acid (ABA). ABA is the major plant hormone related to water stress signaling and regulates plant water balance [122, 123, 124]. Rocha et al., [124] found drouight responses in sugarcane analogous to those induced by exogenous ABA application. Both drought and ABA induced the expression of genes encoding a PP2C- like protein phosphatase, a S-adenosylmethionine decarboxylase and two delta – 12 oleate desaturases. Trujillo et al., [125] showed that SodERF3, a sugarcane ethylene responsive factor (ERF), is induced by ABA and drought stress and may be involved in salt and drought tolerance. However, plant response to drought is a complex phenomenon, especially with a polyploid genome like sugarcane [126]. Phosphorus supply improved sugarcane acclimation capacity by affecting plant characteristics related to water status and photosynthetic performance and causing network modulation under water deficit [127].

6.2. Salinity Stress

SodERF3 is a new member of the FT-ERF family in sugar cane, closely related to salinity and drought tolerance. The C-terminal repression (EAR) motive in *SodERF3* is different from that described to date for other plants transcription factors [128]. Salts interfere with sugar production in two ways: firstly, by affecting the growth rate and cane yield and secondly, by affecting the sucrose content of the stalk [129]. Patade et al., [130] studied the effects of salt and drought stresses on irradiated cells of sugarcane and obtained plants tolerant to higher salt stress. Gandonou et al., [131] studied the effects of salt stress by exposing the callus to a single level of 68 mM NaCl, and observed that physiological and biochemical indicators could play a crucial role in salt tolerance.

6.3. Temperature Extremes

RNA expression profile of sugarcane subjected to low temperature generated reported about twenty novel transcripts/genes [132]. Expression profiles of the cold inducible genes revealed proteins that are directly involved in chilling tolerance. One sugarcane EST encoding a putative xanthine dehydrogenase (XDH) was induced after cold exposure and might be involved in the protection against oxidative stress due to cold exposure [132]. In sugarcane the heat stress effects are reversible through small heat shock proteins (sHsp), which belong to chaperone family [133].

7. Technology Intervention to Overcome Abiotic Stresses

Since climate change is projected to reduce sugarcane yields in the next century, it is vital to come up with mitigation strategies that can lower the effects. A number of mitigation measures can be drawn from understanding the potential effects of climate change relying much on climate models. However, the projections of climate change using models are uncertain because of errors that may be encountered in these models [134]. Water stress generated by high temperatures and low rainfall can be mitigated by growing varieties that are tolerant or resistant to drought.

Inman-Bamber et al., [135] reported that sugarcane cultivar differences in drought adaptation exist. Therefore, evolution of sugarcane varieties with drought resistance and search for genotypes, which possess inherent capabilities of drought tolerance has been on the forefront of sugarcane research. Researchers should therefore continue to breed sugarcane varieties or cultivars that adapt to drought conditions or greater water use efficiency [136]. Hence, each year genetic potential of advanced breeding material (AVT clones) are being exploited for adverse environments like drought and salinity wherein the competitive advantage for one cultivar over another is likely to be greater. Traits such as higher NMC, maintenance of better leaf production, higher single cane weight and rapid stem elongation contribute to cane yield under drought [137].

7.1. Drought

Drought

- Early Planting: In the tropical belt, November to January planting is better than March- April planting to overcome the problem of moisture stress.
- Seed rate and spacing : Higher seed rate (or) closer spacing is to establish a higher stalk population to make up the greater less of individual stalks, row spacing can be narrowed down to 60 (or) 75 cm to give 15- 20 % higher cane yield over normal spacing.
- Seed treatment: Soaking the setts in a saturated lime solution for one hour before planting (dissolve 80 kg of lime (calcium carbonate $CaCO_3$) in 400 lit of water.

- Trash mulching: Trash mulching (5-7 t/ha) helps in conserving soil moisture, checks the weed growth and reduces the soil temperature by 2°C.
- Deep trench system of planting: Deep trench system of planting helps deep root development and efficient use of nutrients and moisture.
- Foliar application of urea and potassium: Foliar application of urea and KCl each at 2.5% (2.5 kg urea + 2.5 kg MOP in 100 liters of water) at 15-20 days interval maintains the crop turgidity.
- Protective irrigation: During drought available water can be given in skipped furrows alternatively.
- Use of anti-transpirants: Kaolin acts as a reflectant and reduces the transpiration loss.
- Tolerant varieties: Varieties CoC 671, Co 8208, Co 85007, Co 85004, Co 86032, Co 85019 and Co 87263, Co 99004 (Damodar), Co 2001-13 (Sulabh) and Co 2001-15 (Mangal), Co 0218,Co 0325 and Co 0328, Co 0403, Co 06015 and Co 06022 are suitable for water limited condition.

7.2. Waterlogging Management

Climate change is projected to result in floods in some areas or years. Since floods result in waterlogging conditions, salinity and raised water table, reducing yields significantly ([81]Glaz et al., 2004), it is therefore important to adapt sugarcane production to such conditions. Drainage systems in the fields that are likely to be affected (flat) areas may need to be installed. Once the drainage improves, excessive salts causing salinity can be leached by irrigation. Varieties that adapt to waterlogging and saline conditions may be grown.

- Proper drainage system should be provided
- Early and deep planting is beneficial.
- Seed rate and row spacing: Normally 38 to 40 thousand healthy three-bud setts/ha are used for planting.
- The row-to-row distances should be widened and deepened to 135 cm to make drainage channels in between them at the time of water logging.
- Split application of nitrogen (2-3 times) helps in minimizing nitrate leaching, the chances of which are more under water-logging.
- Foliar application of 5 per cent urea during water logging increases the yield of cane.
- Foliar application of potassium and phosphorus along with nitrogen causes greater root proliferation and stiffness of cane.
- Sugarcane matures earlier in waterlogged areas, early harvesting facilitates to get maximum amount of sucrose.
- Waterlogging tolerant varieties

Co 8231, Co 8232, Co8145, CoSi 86071, Co Si 776, Co 8371 & Co 99006 by SBI. At Anakapalle, promising clones 93A4, 93A11, 93A145, and 93A21 have been identified under water logging conditions.

In the Kolhapur region of Maharashtra, Co 8371 has been found to perform well under river flood. Some of the Bo varieties like Bo 91 and varieties Co 87263 and Co 87268 are suitable for flooded conditions of Bihar, while CoT1 8201 and CoT1 88322 are grown in Kerala where water logging is very common.

7.3. Salinity Management

In India about one fourth of the acreage is affected by salinity and or alkalinity to varying degrees. Soil salinity and poor quality irrigation water coupled with moisture stress during high water demand period (in summer months), largely coinciding with rapid growth phase of sugarcane results in very low yields. It is estimated that 33% of cane area in Tamil Nadu, 40% of cane area in Andhra Pradesh and 48% in Karnataka experience either soil salinity or saline irrigation water and yield losses reported were about 40 per cent.

- Seed rate: Higher seed rate of 25% is recommended to compensate for germination loss and to ensure adequate crop stand.
- Trench planting: Following "modified trench system" of planting in saline soils and salt water irrigated areas has recorded improved yields of around 15 per cent.
- Use of organic manure: Organic manures viz, pressmud (10-15 t/ha), farmyard manure (25 t/ha), etc., improve the availability of essential (Zn, Fe, Ca and Mn).
- Amendments: With increase in soil pH the requirement of gypsum also varies.
- For application of gypsum soil pH determination is a prerequisite.

Soil pH	Gypsum requirement (t/ha)
8.5-9.0	2.5
9.0-9.5	3.5-7.5
9.6-10.0	8.5-12.5

- Irrigation with good quality water: During critical growth stages (up to 150 days of crop age) irrigation with good quality water is beneficial.
- Mulching: Trash mulching reduces loss of moisture through evaporation thereby minimizing the effect of ions in the soil. Trash upon incorporation adds organic matter and nutrients.
- Green manures: Growing green manure as inter crop and insitu incorporation of green manure highly beneficial to improve productivity - in salt affected soils.
- Nutrient management: 25% Additional nitrogen dosage has been found to improve yield under salinity conditions. Application of top dressings of N and P fertilizer through pocket manuring is advantageous and helps in improving yield significantly.
- Varieties: Among the popular varieties tested, Co 86011, Co 7717, Co 7219, Co 8208, Co 85004, CoC 671, Co 6806,Co 93005, Co 95003, Co 94008, Co 85019, Co 94012, Co 97008 ,Co 99004, Co 2001-13, Co 2001-15, Co 0218, Co 0403 etc., were found suitable for salt affected soils.

Conclusion and Future Research Needs

Extraordinary progress has been made in recent years in diverse areas ranging from meteorological analysis, water balance models which have provided new avenues to achieve improved yields under water limited environments. Using such selection methods, it has been difficult to manipulate precisely the individual components of stress tolerance, and thus development of stress tolerant cultivars has not made much headway. Improvement of qualitative traits such as photosynthesis, transpiration, roots and water use efficiency is difficult through conventional breeding because of the complexity of quantifying these traits in a large number of germplasm lines. Therefore, the immediate task ahead is to look for new options to exploit the variability present in our rich germplasm through new evaluation techniques. Several stress related genes that alter the expression of important stress traits for improving the plant adaptation have to be fully characterized. The donor sources for various stress resistant features have to be identified and the genes responsible for these traits need to be transferred through molecular breeding. In recent years, many genes and gene products have been identified which get induced upon exposure to various abiotic stresses [137]. Genes encoding enzymes of the biosynthetic pathways of different osmolytes such as proline, glycine betaine, sorbitol, pinitol, have been cloned and exploited in improving abiotic stress tolerance. Response of crop plants to abiotic stresses involves several gene alterations. Proteins which are up regulated are synthesized in response to high temperature, drought, salinity and several other abiotic stresses. New research programme should be aimed at identifying or developing cultivars and management practices appropriate for altered climates. Besides breeding new varieties of sugarcane to mitigate effects of climate change, scientists can also use biotechnology to reduce abiotic and biotic stresses associated with sugarcane. Genetically modified sugarcane is expected to have a multiple tolerance potential viz., increasing yield, drought tolerance and insect resistance of sugarcane. Biotechnology also releases varieties faster than conventional breeding of sugarcane.

References

[1] Dwivedi RS (2000) Adaptability mechanisms of sugarcane cultivars to abiotic stresses. In Sugarcane Production: Strategies and Technology (Shahi HN, Lal M, Sinha OK, Srivastava TK, eds), Technical Bulletin No. 40, Indian Institute of Sugarcane Research, Lucknow, pp 42-45.

[2] Zende NA and Hapase DG (1986) Development of saline and alkaline soils in Maharashtra under sugarcane cultivation, their reclamation and mamgement. *Bahratiya Sugar* 11(11):41-48.

[3] Zende GK (2002) Soil water management in waterlogged soils. *Bharatiya Sugar* 27(5): 7-8.

[4] Boyer JS (1982). Plant productivity and environment (crop genetic improvement). *Science.*218 (4571: 443-448.

[5] Jones HG and Corlett JE (1992) Current topics in drought physiology. *Journal of Agricultural Science* 119:291-296.

[6] Sundara B (1998) Sugarcane Cultivation, Vikash Pub. House, New Delhi, pp. 292.

[7] Naidu KM (1987) Potential yield in sugarcane and its realization through varietal improvement- present status and future thrusts. Sreenivasan, T.V. and Premachandran, M.N. (eds.)., *Sugarcane Breeding Institute*, pp 19-55.

[8] Naidu KM and Venkataramana (1993) Sugarcane In "Rooting pattern of tropical crops" Ed. M.A. Salam, Tata McGraw Hill (India) New Delhi pp. 169-187.

[9] Venkataramana S and Naidu KM (1989) Root growth during formative phase in irrigated and water stressed sugarcane and its relationship with shoot development and yield. *Indian J. Plant Physiol.* 32: 43-50.

[10] Sheu YS and Yang PC (1983). A preliminary study on agronomic characters in relations to solar utilization in sugarcane. *Taiwan Sugar*, 39: 124-128.

[11] Shih SF and Gascho GJ (1980). Sugarcane stalk distribution in two row spacing. *Int. Soc. Sug. Tech. Cong.* Manila, Proc 17: 38-51.

[12] Mongelrad JC (1968) Further notes on the use of Sin bar for the selection of drought resistant sugarcane varieties. *Annl. Rept. M S R I Mauritius*: 87-89.

[13] Rao PN (2000) Cane management under drought conditions or water stress. *Proc. Annl. Conv. S.T.A.I.* 61: 3-8.

[14] Mongelrad JC (1968a) Effects of planting methods on yield of sugarcane under sub-surface irrigation. *Hawaiian Planter's Rec.* 58(22):315-322.

[15] Bull TA and Glasziou KT (1975) Sugarcane. In Crop Physiology, case histories, Ed. LT Evans, Cambridge Univ. Press, Cambridge, pp 51-72.

[16] Almazan O, Gonzalez L and Galvez L (2001) Sugar Cane International, 7, 3-8. Journal of Agricultural Science, Cambridge (1992), pp 119, 291-296. 1992 Cambridge University Press 291.

[17] Luo Jun, Zhang MuQuing, Lu JianLin and Lin YanQuan (1999) Chloroplast fluorescence parameters, MDA content and plasma membrane permeability in sugarcane and their relation to drought tolerance. *J. Fujian Agricultural University* 28 (3): 257-262.

[18] Luo Jun, Zhang MuQuing, Lu JianLin and Lin YanQuan (2000) Effects of water stress on the chlorophyll a fluorescence induction kinetics of sugarcane genotypes. *J. Fujian Agricultural University* 29 (1):18-22.

[19] Naidu KM, Ramana Rao TC and Shunmugasundaram S (1983a) Varietal stability in sugarcane under conditions of drought. *Proc. Int. Soc. Sugarcane Technol.,* 18: 682-690.

[20] Venkataramana S, Gururaja Rao PN and Naidu KM (1986) The effects of water stress during the formative phase on stomatal resistance and leaf water potential and its relationship with yield in ten sugarcane varieties. *Field Crops Res.* 13:345-353.

[21] Gururaja Rao, PadmajaRao S and Venkataramana S (2008) *Annual Report*, SBI, Coimbatore p.

[22] Ramanujam T and Venkataramana S (1999) Radiation interception and utilization at different growth stages of sugarcane and their influence on yield. *Indian J. Plant Physiol.* 4: 85-89.

[23] Venkataramana S, Shunmugasundaram S and Naidu KM (1984) Growth behaviour of field grown sugarcane varieties in relation to environmental parameters and soil moisture stress.*Agric. & For. Meteorol.* 31: 251-260.

[24] Naidu KM and Venkataramana S (1989) Harvest Index in water stressed sugarcane varieties *Sugarcane* (London) No. 6: 5-7.

[25] Venkataramana S, Naidu KM and Singh M (1987) Membrane thermostability and nitrate reductase activity in relation to water stress tolerance of young sugarcane plants. *New Phytol.*107 (2): 335-340.
[26] Venkataramana S, Gururaja Rao PN and Naidu KM (1983) Evaluation of cellular membrane thermostability for screening drought resistant sugarcane varieties, *Sugarcane* (London) No. 4: 13-15.
[27] Rao KC and Asokan S (1978) Studies on free proline association to drought resistance in sugarcane. *Sugar J.*, 40: 23-24.
[28] Vasantha S and Gururaja Rao (2003) Influence of moisture stress on the activity of oxidative enzymes in sugarcane *India Journal Plant Physiol.* 8: 405-407.
[29] Carlin SD and Santos DMM (2009) Physiological indicators of the interaction between water deficit and soil acidity in sugarcane. *Pesquisa Agropecuária Brasileira* 44: 1106-1113.
[30] Queiroz RJB, Durvalina Maria Mathias dos Santos, Antonio Sergio,Ferraudo,, Samira Domingues Carlin and Marcelo de Almeida Silva. (2011). Biochemical and physiological responses of sugarcane cultivars to soil water deficiencies. *Sci. Agric.* 68(4): 469-476.
[31] Ramesh P (2000) Effect of drought on nutrient utilization, yield and quality of sugarcane *(Saccharrum officinaram)* varieties. *Indian J.Agron.* 45(2): 401-406.
[32] Samuels G (1971) Influence of water deficiency and excess on growth and leaf nutrient element content of sugarcane. *Proc. International Soc. Sugar Cane Technol.* 14: 653-656.
[33] Venkataramana S and Naidu KM (1993) Abscisic acid effect on water stress indicators in sugarcane *Plant Physiol. & Biochem.* 20: 1-4.
[34] Vasantha, Gururaja Rao PN and Ramanujam T (2003) Effect of moisture stress on sucrose synthesizing enzymes in promising genotypes of sugarcane *J. Plant Biol.*30: 15-18.
[35] Gascho GJ and Shih SF (1983) Sugarcane. In Crop-Water Relations, Ed Teare ID and MM Peet, Wiley Inter science, New York, 445-479.
[36] Chang H, Wang JS and Ho FW (1968) The effect of different pan ratio for controlling irrigation of sugarcane in Taiwan. *Proc. Int. Soc. Sugarcane Technol.* 13: 652-663.
[37] Shih SF and Gascho GJ (1980) Relationship among stalk elongation, leaf area and dry biomass of sugarcane. *Agron. J.* 72: 309-313.
[38] Munns R (2002) Comparative physiology of salt and water stress. *Plant Cell Environ.* 25, 239-250.
[39] Maas EV (1986) Salt tolerance of plants. *Applied Agricultural Research* 1**:** 12-26.
[40] Plaut Z, Meinzer F and Federman E (2000) Leaf development, transpiration and ion uptake and distribution in sugarcane cultivars grown under salinity. *Plant Soil* 218: 59-69.
[41] Kumar S, Naidu KM and Sehtiya HL (1994) Causes of growth reduction in elongating and expanding leaf tissues of sugarcane under saline conditions. *Australian Journal of Plant Physiology*21: 79-83.
[42] Liu LJ (1967) Salinity effects on sugarcane germination, growth and root development. *J.Agric. Univ. P.R.* 51: 201-209.

[43] Chowdhury MKA, Miah MAS, Ali S, Hossain MA and Alam Z (2001) Influence of Sodium chloride salinity on germination and growth of sugarcane (*Saccharum officinarum* L.) *Sugarcane International* : 15-16.

[44] Kumar S and Naidu KM (1993) Germination of sugarcane setts under saline conditions. *Sugarcane* 4: 2-5.

[45] Syed MM and El- Swaify SA (1972) Effect of saline water irrigation on NCO 310 and H 50-209 cultivars of sugarcane. *Tropical Agri.* 49(4): 337-346.

[46] Joshi GV and Naik GR (1977) Salinity effect on growth and photosynthetic productivity in sugarcane variety Co 740. *Indian Sugar* 27(6): 329-332.

[47] Joshi GV and Naik GR (1980) Response of sugarcane to different types of salt stress. *Plant and Soil* 56(2): 255-263.

[48] Naik GR and Joshi GV (1981) Effect of pre planting treatment with growth substances on sugarcane productivity under saline conditions. *Deccan Sugar Technologists Association.31st Annual Convention*.A77-86.

[49] Meinzer FC, Palut Z and Saliendra NZ (1994) Carbon isotope discrimination, - exchange and growth of sugarcane cultivars under salinity. *Plant Physiology* 104: 521-526.

[50] Wahid A (2004) Analysis of toxic and osmotic effects of sodium chloride on leaf growth and economic yield of sugarcane. *Bot. Bull. Acad. Sin.* 45: 133–41.

[51] Sharma SK, Sharma P and Upal SK (1997) Influence of salt stress on growth and quality of sugarcane. *Indian Journal of Plant Physiology* 2:179-180.

[52] Gomathi R and Thandapani P (2004) Influence of salt stress on yield and quality of sugarcane genotypes (*Saccharum officinarum* L).*Indian Sugar*. 117-124.

[53] Vasantha S (2003) Varieties for salt affected soils. 35th meeting of Sugarcane Research and Development workers of Tamil Nadu. Sep 8-9, 2003. Agenda notes pp 25-29.

[54] Gomathi R and Thandapani V (2005) Salt stress in relation to nutrient accumulation and quality of sugarcane genotypes. *Sugar Tech*. 7(1): 39-47.

[55] Vasantha S, Gomathi R and Rakkiyappan P (2009) Sodium content juice and jaggery quality of sugarcane genotypes under salinity. *Journal of Biological Sciences* 1(1) :33-38.

[56] Gomathi R, Vasantha S, Thandapani V and Thandapani P (2010) Mechanism of Osmo - Regulation in Response to Salinity Stress in Sugarcane. *Sugar Tech*. 12(3–4):305–311.

[57] Vasantha S, Venkataramana S, Gururaja Rao PN and Gomathi R (2010) Long term salinity effect on growth, photosynthesis and osmotic characteristics in sugarcane. *Sugar Tech*. 12(1) 5-8.

[58] Vasantha S, Gururaja Rao PN, Venkataramana S and Gomathi R (2008) Salinity induced changes in the antioxidant response of sugarcane genotypes. *Journal of Plant Biology* 35 (2):115-117.

[59] Gomathi R and Rakkiyapan P (2011) Comparative lipid peroxidation and leaf membrane thermostability, antioxidant system in four sugarcane genotypes differing in salt tolerance. *Int. J. Plant Physiol. & Biochem*. 3(4):51- 58.

[60] Vasantha S and Rajalakshmi S (2009) Progressive changes in biochemical characters in sugarcane genotypes subjected to NaCl treatment. *Indian Journal of Plant Physiology* 14(1):34-38.

[61] Gomathi R, Yukashini K, Shiyamala S, Vasantha S, Suganya A and Rakkiyappan P (2013). Induced Response of Sugarcane Variety Co 86032 for Thermotolerance. *Sugar Tech.* 15(1):17–26.

[62] Carter CE and Floyed JM (1974) Inhibition of sugarcane yields by high water level during dormant season. *Proc. Soc. Sugarcane Technol.* 4:14-18.

[63] Carter CE (1976) Excess water decreases cane and sugar yields. *Proc. American Soc. Sugar cane Technol.* 6:44-51.

[64] Barcly AM and Crawford RMM (1982) Plant growth and survival under strict anaerobiosis. *J. Expt. Bot.* 33:541-549.

[65] Gomathi R and Chandran K (2009) Effect of waterlogging on growth and yield of sugarcane clones. Sugarcane Breeding Institute (SBI-ICAR), *Quarterly News letter* Vol.29 (4).

[66] Mc Mahon GG, Chapple PA, Ham GJ, Saunders M and Brandon R (1993) Planting sugarcane on heavy clay soils in the Burdekin.Proc.15th conference of the Australian Society of Sugarcane technologist, Cairns, Queensland, 27-30.

[67] Glaz B (2003) Sugarcane genotypes emergence after rain-caused flood. *Sugarcane* International.Issue.35

[68] Jackson MB and Richard B (2003) Physiology, biochemistry and molecular biology of plant root systems subjected to flooding of the soil. In H. de Kroon and E. J. W. Visser /9eds.), Root ecology. Springer – Verlag, Berlin, Heidelberg

[69] Van der Heyden, Ray CJ and Noble R (1998) Effects of waterlogging on young sugarcane plants. *Australian Sugarcane 2:* 28 – 30.

[70] Sukchain DS and Dhaliwal LS (2005) Correlations and path coefficients analysis for aerial roots and various other traits in sugarcane under flooding. *Ann. of Bio.* 21: 43 – 46.

[71] Drew MC (1997) Oxygen deficiency and root metabolism: Injury and acclimation under hypoxia and anoxia. *Annual Review of Plant Physiology Plant Molecular Biology* 48: 223-250.

[72] Drew MC and Sisworo ET (1979) The development of waterlogging damage in young barley plants in relation to plant nutrient status and changes in soil properties. *New Phytologist* 82, 301–314.

[73] Jackson MB (1990) Communication between the roots and shoots of flooded plants. In: Importance of root to shoot communication in the responses to environmental stress (Davies, W.S. and Geffocat, B., Eds.), *British Society for Plant Growth Regulation Bristol.* U.K. pp. 115-133.

[74] Sarkar RK, Sahu RK, Suriya Rao AV, De RN (1999) Correlation and path analysis of certain morpho-physiological characters with submergence tolerance in Rain - fed lowland rice. *Indian Journal Plant Physiology* 4(4): 346-348.

[75] Gomathi R (2009) Physiological basis of waterlogging resistance in sugarcane. In. Sugarcane Physiology. Ed. By. P.N. Gururaja Rao, Published by Sugarcane Breeding Institute (ICAR) and Directorate of open and distance learning, TNAU, Coimbatore, Tamil Nadu.pg.53-61.

[76] Webster PWD, Eavis BW (1971) Effect of flooding on sugarcane growth 1.Stage of growth duration of flooding. *Proc. of Int. Soc. Sugarcane Tech.* 14:708-714.

[77] Rahman ABMM, Martein FA, Terry ME (1989) Physiological response of sugarcane to flooding stress. *Proc. Int. Soc. Sugar. Tech.* 20(2): 668-676.

[78] Gomathi R Chandran K (2008) *Annual Report*, SBI, Coimbatore 39-40.

[79] Pandey OP, Asha Kumari, Haque H (2000) NAD-Alcohol dehydrogenase and Superoxide Dismutase activity in *Zea mays* under Hypoxia and post-Hypoxia stress Regime. *Journal of Plant Biology27*(1): 71-73.

[80] Manoharan ML, Duraisamy K, Krishnamurthy SV, Vijayaraghan H, Muthkrishnan K (1990) Performance sugarcane varieties in waterlogged condition. *Maharastra Sugar.*15(11): 39-45.

[81] Giaz B, Morris DR, Daroub SH (2004) Sugarcane photosynthesis, transpiration and stomatal conductance due to flooding and water table. *Crop Sci.* 44: 1633 – 1641.

[82] Giaz B, Gilbert RA (2006) Sugarcane response to water table, periodic flood, and foliar nitrogen on organic soil. *Agronomy Journal* 98: 616 – 621.

[83] Sachs MM, Freelins M, Okimoto R (1980) The anaerobic proteins of maize. *Cell*20: 761-767.

[84] Mujer CV, Rumpho ME, Lin JJ, Kennedy RA (1993) Constitute and inducible aerobic and anaerobic stress proteins in the *Echinochloa* complex and rice. *Plant Physiology*101: 217-226.

[85] Lin CH, Lin CH (1992) Physiological adaptation of wax apple to waterlogging. *Plant Cell Environment* 15: 321-328.

[86] Liao CT, Lin CH (1995) Effect of flood stress on morphology and anaerobic metabolism of Momordicacharantia. *Environmental Experimental Botany 35:* 105-113.

[87] Russell DA, Wong DML, Sachs MM (1990) The anaerobic response of soybean. *Plant Physiology* 92: 401-407.

[88] Kennedy RA, Rumpho ME, Fox TC (1992) Anaerobic metabolism in plants. *Plant Physiology*100: 1-6.

[89] Gomathi R, Gowri Manohari, Rakkiyappan P (2012) Antioxidant enzymes on cell membrane integrity of sugarcane varieties differing in flooding Tolerance. *Sugar Tech*14*:* 261-265.

[90] Weijun Zhou, Xianqing Lin (1995) Effects of waterlogging at different growth stages on physiological characteristics and seed yield of winter rape *(Brassica napus L.). Field Crops Research* 44:103-110.

[91] Ashraf M, Hafeez M (2004) Thermotolerance of pearl millet and maize at early growth stages: growth and nutrient relations. *Biol. Plant*. 48: 81–86.

[92] Wahid A (2007) Physiological implications of metabolites biosynthesis in net Assimilation and heat stress tolerance of sugarcane sprouts. *J. Plant Res.* 120:219–228.

[93] Hall AE (1992) Breeding for heat tolerance. *PlantBreed.Rev.*10, 129– 168.

[94] Ebrahim MK, Zingsheim O, El-Shourbagy MN, Moore PH, Komor, E (1998) Growth and sugar storage in sugarcane grown at temperature below and above optimum. *J. Plant Physiol.* 153: 593–602.

[95] Adams SR, Cockshull, KE and Cave, CRJ (2001) Effect of temperature on the growth and development of tomato fruits. *Ann. Bot*. 88: 869–877.

[96] Vinocur B, Altman A (2005) Recent advances in engineering plant tolerance to abiotic stress: achievements and limitations. *Curr. Opin. Biotechnol*. 16, 123–132.

[97] Bohnert HJ, Gong Q, Li P, Ma S (2006) Unraveling abiotic stress tolerance Mechanisms-getting genomics going. *Curr. Opin. Plant Biol*. 9: 180–188.

[98] Gomathi, R, Shiyamala S, Vasantha S and Suganya A (2014). Optimization of temperature conditions for screening thermotolerance in sugarcane through temperature induction response (TIR) technique. *International J. Science*. 3(2): 5-18.

[99] Wang W, Vinocur B, Shoseyov O, Altman A (2004) Role of plant heat-shock Proteins and molecular chaperones in the abiotic stress response. *Trends Plant Sci*. 9:244–252.

[100] Vu JCV, Allen LH, Boote KJ, and Bowes G (1997) Effect of elevated CO_2 pump. *Photosynthetic Res*. 77.191-207.

[101] Fuhrer J (2003). Agroecosystem response to combination of elevated Co2, ozone and global climate change. *Agric. Ecosyst.Environ*.97:1-20.

[102] Rosenberg NJ, Blad BL and Verma SB (1983) Microclimate: The biologica Environment. Jojn Wiley and Sons, new York, 495 pp.

[103] Taiz L and Zeeige E (1991) Plant Physiology. The Benjamin/ Cummings, New York, p.559.

[104] Wahid A and Ghazanfar A (2006) Possible involvement of some secondary metabolites in salt tolerance of sugarcane. *J. Plant Physiol*. 163:723–730.

[105] Simoes-Araujo JL, Rumjanek NG and Margis-Pinheiro M (2003) Small heat shock proteins genes are differentially expressed in distinct varieties of common bean. *Braz. J. Pl. Physiol*. 15: 33–41.

[106] Wahid A and Close TJ (2007) Expression of dehydrins under heat stress and their relationship with water relations of sugarcane leaves. *Biol. Plant*. 51:104–109.

[107] Chalker-Scott L (2002) Do anthocyanins function as osmoregulators in leaf tissues? *Adv. Bot. Res*. 37, 103–106.

[108] Jain Radha, Shrivastava AK, Solomon S and Yadav RL (2007) Low temperature stress-induced biochemical changes affect stubble bud sprouting in sugarcane (*Saccharum spp*. hybrid). *Plant Growth Regul*. 53:17–23.

[109] Hartt CE (1965) Effect of temperature upon translocation of C^{14} in sugarcane. *Plant Physiol*. 40: 74-81.

[110] Inman-Bamber NG, Bonnett GD, Spillman MF, Hewitt MH and Glassop D (2010) Sucrose accumulation in sugarcane is influenced by temperature and genotype through the carbon source sink balance. *Crop Pasture Sci*. 61: 111-121

[111] Du YC and Nose A (2002) Effects of chilling temperature on the activity of enzymes of sucrose synthesis and the accumulation of saccharides in leaves of three sugarcane cultivars differing in cold sensitivity. *Photosynthetica* 40:389-395.

[112] Bray EA (1997) Plant responses to water deficit *Trends in Plant Science* 2: 48-54.

[113] Thomashow MF (1999) Plant cold acclimation: freezing tolerance genes and regulatory mechanisms. Annu. Rev. Plant Physiol. *Plant Mol. Biol*. 50**:** 571–599.

[114] Shinozaki K, Yamaguchi-Shinozaki K and Seki M (2003) Regulatory network of gene expression in the drought and cold stress responses. *Curr. Opin. Plant Biol*. 6: 410–417.

[115] Cushman JC and Bohnert HJ (2000) Genomic approaches to plant stress tolerance. *Curr. Opin. Plant Biol*, 3117-124.

[116] Rodrigues FA, Laia ML and Zingaretti SM (2009) Analysis of gene expression profiles under water stress in tolerant and sensitive sugarcane plants. *Plant Sci*. 176: 286-302.

[117] Prabhu G, Kawar PG, Pagariya MC and Prasad DT (2011) Identification of water deficit stress upregulated genes in sugarcane. *Plant Mol. Biol. Rep*. 29.

[118] Chagas RM, Silveira JAG, Ribeiro RV, Vitorello VA and Carrer H (2008) Photochemical damage and comparative performance of super oxide dismutase and

ascorbate peroxidase in sugarcane leaves exposed to paraquat-induced oxidative stress. *Pestic. Biochem. Physiol.* 90:181-188.

[119] Zhang SZ, Yang BP, Feng CL, Chen RK, Luo JP, Cai WW and Liu FH (2006) Expression of the *Grifola frondosa* trehalose synthetase gene and improvement of drought-tolerance in sugarcane (*Saccharum officinarum* L.). *J. Integr. Plant Biol.* 48:453-459.

[120] Molinari HBC, Marur CJ, Daros E, Campos MKF, Carvalho JFRP, Bespalhok Filho JC, Pereira LFP and Vieira LGE (2007) Evaluation of the stress-inducible production of proline in transgenic sugarcane (*Saccharum* spp.): osmotic adjustment, chlorophyll fluorescence and oxidative stress. *Physiol. Plant.* 130: 218-229.

[121] Guimaraes ER, Mutton MA, Mutton MJR, Ferro MIT, Ravaneli GC and Silva JA (2008) Free proline accumulation in sugarcane under water restriction and spittlebug infestation. *Sci. Agric.* 65: 628 – 633.

[122] Riera M, Valon C, Fenzi F, Giraudat J and Leung J (2005) The genetics of adaptive responses to drought stress: abscisic acid-dependent and abscisic acid-independent signaling components. *Physiol. Plant.* 123: 111-119.

[123] Wilkinson S and Davies WJ (2010) Drought, ozone, ABA and ethylene: new insights from cell to plant community. *Plant Cell Environ.* 33:510-525.

[124] Rocha FR, Papini-Terzi FS, Nishiyama MY Jr, Vencio RZN, Vicentim R, Duarte RDC, Rosa VE Jr. Vinagre F, Barsalobres C, Medeiros AH, Rodrigues FA, Ulian EC, Zingaretti SM, Galbiatti JA, Almeida RS, Figueira AVO, Hemerly AS, Silva-Filho MC, Menossi M and Souza GM (2007) Signal transduction-related responses to phytohormones and environmental challenges in sugarcane. BMC Genomics 8:71.

[125] Trujillo LE, Sotolongo M, Menendez C, Ochogavia ME, Coll Y, Hernandez I, Borras-Hidalgo O, Thomma BPHJ, Vera P and Hernandez L (2008) SodERF3, a novel sugarcane ethylene responsive factor (ERF), enhances salt and drought tolerance when overexpressed in tobacco plants. *Plant Cell Physiol.* 49:512-525.

[126] Grivet L and Arruda P (2001) Sugarcane genomes: depicting the complex genome of an important tropical crop. *Curr. Opin. Plant Biol.* 5: 122-127.

[127] Sato AM, Catuchi TA, Ribeiro RV and Souza GM (2010) The use of network analysis to uncover homeostatic responses of a drought-tolerant sugarcane cultivar under severe water deficit and phosphorous supply. *Acta. Physiol. Plant.* 32:1145-1151.

[128] Trujillo LE, Carmen Menéndez, María E Ochogavía, Ingrid Hernández, Orlando Borrás, Raisa Rodríguez, Yamilet Coll, Juan G Arrieta, Alexander Banguela and Ricardo Ramírez, Lázaro Hernández (2009) Engineering drought and salt tolerance in plants using SodERF3, a novel sugarcane ethylene responsive factor *Biotecnología Aplicada* 26(2): 168-171.

[129] Rozeff N (1995) Sugarcane and salinity-a review paper. *Sugarcane* 5:8-19.

[130] Patade VY, Suprasanna P, Kulkarni UG and Bapat VA (2006) Molecular profiling using RAPD technique of abiotic stress (salt and drought) tolerant regenerants of sugarcane Cv. Coc-671. *Sugar Tech.* 8: 63-68.

[131] Gandonou Ch, Errabii T, Abrini J, Idaomar M and Skali Senhaji N. (2006). Selection of callus cultures of sugarcane (Saccharum sp.) tolerant to NaCl and their response to salt stress. *Plant Cell, Tissue and Organ Culture*, 87: 9-16.

[132] Nogueira FTS, De Roza Jr VE, Menossi M, Uhan EC and Arruda P (2003) RNA Expression profiles and data mining of sugarcane response to low temperature. *Plant Physiol.* 132: 1811-1824.

[133] Tiroli-Cepeda AO and Ramos CHI (2010) Heat causes oligomeric disassembly and increases the chaperone activity of smaller heat shock proteins from sugarcane. *Plant Physiol. Biochem*. 48:108-116.

[134] Mall RK, Lal M, Bhatia VS, Rathore LS and Singh R (2004) Mitigating climate change impact on soybean productivity in India: a simulation study. *Agric. For. Meteorol.* 121(1–2):113–125.

[135] Inman-Bamber N G, Lakshmanan P and Park S (2012) Sugarcane for water-limited environments: Theoretical assessment of suitable traits. *Field Crops Res*. 134:95–104.

[136] Mathieson L (2007) Climate change and the Australian Sugar Industry: Impacts, adaptation and R & D opportunities. *Sugar Research and Development Corporation.* Australia.

[137] Chaves MM, Maroco JP and Pereira JS (2003) Understanding plant responses to drought – from genes to the whole plant. *Functional Pl. Biol*. 30:239-264.

In: Current Status of Sugarcane Research in India
Editors: A. K. Tiwari, M. Lal and A. K. Singh
ISBN: 978-1-63463-458-8

Chapter 5

SUGARCANE MICROPROPAGATION FOR QUALITY SEED PRODUCTION AND CONSTRAINTS FOR MASS ADAPTABILITY

S. G. Dalvi*

Tissue Culture Section,Vasantdada Sugar Institute, Pune, India

ABSTRACT

Sugar industry seen as a multi product industrial complex, manufactures not only sugar but also industrial /potable alcohol, fuel ethanol, electricity and bio-gases as an equally valuable activity. Sugar manufacturing, besides its contribution to GDP is also a source of substantial rural/semi-rural employment. Further it can potentially contribute to renewable energy pool in a significant manner making sugarcane as future renewable energy crop. Looking in to future demand of sugar for consumption, sugarcane for cogeneration and ethanol production, many new sugar units have recently been erected/are in erection phase. The limited cultivable area available for expansion and continuing conversion of agricultural land for non-agricultural purposes necessitate that increase in sugarcane production comes mainly from increase in per hectare yield. Being vegetatively propagated and long duration crop sugarcane is faced with different abiotic biotic stresses resulting in degeneration of varieties. Most of the sugarcane farmers in India are confronted with problems of low cane yield due to poor quality seed and several other local constraints.

Quality seed production through three tier seed system implemented through conventional Moist Hot Air Treatment has many limitations. Micropropagation technology for rapid multiplication of disease-free planting material has greatly facilitated mass production of quality seed in sugarcane. Studies have further shown that the micropropagation-based sugarcane has significantly better germination, tillering, cane yield, and sugar content than the conventionally raised sugarcane. But this potential has not been fully exploited in our country. The scientists / breeders, sugar mills and the sugarcane farmers, who are the major stakeholders in the value chain, cite different reasons for the absence of credible seed program. Intensive "Lab to Land planning" as

* Corresponding author Email: sg.dalvi@vsisugar.org.in.

well as implementation policy is therefore urgently needed to strengthen the mass adaptability of sugarcane micropropagation technology to sustain sugarcane cropping as well as sugar industry of the country.

Keywords: Micropropagation, Three Tier Seed, Quality Seed Constraints, Sugarcane Seed Policy

INTRODUCTION

Seed quality assumes a great significance in any crop and is the basic and important input in sugarcane cultivation to increase the productivity by 12 to 15 %. The good quality seed material is not used even by 10 percent of total sugarcane farmers. Conventional seed production has many limitations. Therefore, quality seed program through sugarcane micropropagation needs to be adapted systematically so that sugarcane and sugar production become sustainable. In a typical sugar mill, 100 tonnes of sugarcane on an average produce 10 tonnes of sugar, 4 tonnes of molasses from which ethanol is produced, 3 tonnes of press mud which is converted into biofertilizer, 30 tonnes of bagasse used for cogeneration of power to generate 1,500 Kw electricity and for manufacturing paper. Apart from these, about 4 tonnes of cane tops and leaves are generally left in the field, which through recycling further add to the economic value of the crop. Globally there is high demand for sugarcane as a renewable biofuel. Sugarcane, thus, plays a major role in the economy of sugarcane growing regions of the World and hence, improving sugarcane productivity will greatly help in economic prosperity of the farmers and other stakeholders associated with sugarcane cultivation.

In view of the existing resource availability and demand, efficiency-mediated improvement in sugarcane productivity is the most viable option to enhance production and profitability. Sugarcane being a vegetatively propagated crop, there is a need to change the seed after every four years to maintain the purity of the variety and eliminate the diseases and pests being transmitted through setts [2, 3, 7, 8 and 11]. Supply of the setts to farmers for sugarcane plantation is the practice in most of the sugar mill areas with three tier seed program: breeder seed, foundation seed and certified seed. For planting breeder's seed nursery, the moist hot air therapy (MHAT) treated planting material is utilized to get rid of pathogens present in the sugarcane setts. But this treatment results in about 50% reduction in bud germination, which ultimately results in reduced multiplication ratio. In the conventionally planted (MHAT Treated) seed nursery, sett-borne pathogens which cause grassy shoot disease are totally eradicated however whip smut and viruses cannot be totally eradicated [2,3,7,8 and 11]. On the contrary, in sugarcane micropropagation, quality of planting material is excellent as all pathogens including viruses are totally removed as well as the seed multiplication is very fast and economical. Therefore use of tissue cultured plantlets (TCP's) as breeder seed is the best alternative in sugarcane seed production program which ensures that quality seed can be provided very efficiently to large number of sugarcane growers. Looking in to these advantages, micropropagation protocols for different sugarcane varieties and field evaluation for higher seed ratio have been worked out by different tissue culture laboratories and state agriculture universities (2-25). It has been demonstrated that there is improvement in sugarcane yield by 25-30 t/ha and in sugar recovery by 10 -15% with this technology.Therefore, Central Government and State Governments have provided lot of subsidies for the establishment sugarcane micropropagation laboratories for production of tissue cultured plantlets and to sugarcane growers to use micropropagated sugarcane plantlets in three tier seed chain. Though the technology has been well refined and caters to the need of sugar industry; compared to horticultural crops, the speed of adaption of tissue cultured sugarcane plantlets at grass root level farmers is very slow. In this chapter, the factors associated with less adaptability and strategies to increase utilization of micropropagation technique in sugarcane seed production program are discussed.

Sugarcane micropropagation is now becoming a routine technique in sugar industry. A systematic protocol for the production and supply of quality seed in sugarcane using micropropagation technology has been recommended by the Dept. of Biotechnology, Govt. of India (26, 27). For sugarcane micropropagation the apical meristems should be utilized for culture initiation. Mass multiplication of the cultures should be strictly followed as per the standard protocol. For obtaining meristems, shoots tops should be taken from specially maintained Nucleus Seed Nursery. For raising nucleus seed nursery the setts must be obtained from authentic source and should be planted only after the moist hot air treatment. The shoot tops for explants should be taken from vigorously growing disease and pest free tillers of 3-4 month age. The mother culture thus raised should be tested for disease freeness and genetic fidelity using culture indexing techniques for fungal and bacterial microbes as well as with ELISA/PCR/molecular markers for phytoplasma and viruses [2, 11 and 26]. The sub culturing for shoot induction should be followed up to 6-7 cycles, 1-2 root induction cycles and well rooted plants should be transplanted in vector proof green house for acclimatization. The well acclimatized plantlets should be distributed as breeder seed for plantation of foundation seed plots. The foundation and certified seed nurseries raised from the breeder seed should also be verified by the concerned experts for secondary disease and pest infestation as well as purity of seed.

Use of tissue cultured plantlets (TCPs) for raising seed nursery is becoming slowly popular in Maharashtra and adjoining states viz. Gujarat, Madhya Pradesh and Goa because introduction of TCP's in seed chain has many advantages as-

- Hundred percent survival of TCP's in field
- Fast coverage under newly released varieties
- Rejuvenation of old popular varieties
- Assured genetic purity and uniformity
- Absolute disease and pest freeness.
- High seed multiplication ratio under field condition (1:20-30)
- High germination percentage (>95%) in the setts produced using TCP's.
- Early germination of setts (within 8-10 days)
- Useful for spread of only specific varieties in area of sugar mill.
- Increment in yield of commercial crop by 20-25%.
- Uniform cane maturity resulting recovery improvement to 10%.
- Better multiple ratoons.
- Increased availability of quality cane in mills area of operation.
- Reduction in cane transportation cost and increase in mill efficiency.

- Beneficial microbes can be efficiently, effectively and economically introduced through this system.

Due to the above benefits, few sugar mills and progressive sugarcane growers are opting for the tissue cultured seed and there is increasing demand for the TCP's by the sugar mills. However,considering the total area under sugarcane cultivation, the seed replacement through this system is very small. For mass popularizing the use of tissue culture seed, systematic planning should be implemented so that the technology adoption becomes properly and both sugarcane farmers and sugar mills get benefitted.

TCP's as Breeder Seed

Experiences at the Vasantdada Sugar Institute, Pune during the past 10 years shows that the inclusion of tissue cultured plantlets as compared to the conventional system in sugarcane seed chain is quite profitable (Table 1). If the performance of conventional sugarcane planting (MHAT treated seed) and tissue culture system is compared, it is observed that large area is covered with quality seed under tissue culture system within a short period very cost effectively.

Table 1. Comparative Area Coverage by Conventional and TCP raised Seed in Sugarcane

Sr. No.	Production Stage	Conventional (Without MHAT)	Conventional (With MHAT)	TCP's as Breeders Seed
1	Breeder's Seed Nursery	1 ha	1 ha	1 ha (12,500 TCP's*)
2	Foundation Seed Nursery	10 ha	5 ha	20 ha
3	Certified Seed Nursery	100 ha	50 ha	300 ha
4	Commercial Crop	1000 ha	500 ha	3000 ha

* Spacing 3' X 2'.

Thus at every stage in the three tier system of seed multiplication, higher net profit could be obtained with tissue culture system than other systems of seed multiplication. The net overall profit inclusive of all stages is about 10.25 times higher with tissue culture system than in the conventional system. This is because of early and higher germination percentage (90-95%). It also has been demonstrated that in tissue culture seed there is 10% improvement in recovery [2.3, 11 and 26]. Further there are additional benefits in ratoons crops. Due to higher germination and rejuvenation of the seed quality, multiple ratoons can be taken with good economic returns. It is therefore recommended that TCP's should be integrated as a routine practice in the three-tier system of seed multiplication. It has been observed that practically farmers need only 100 TCP's to raise seed sufficient for planting 1 acre area in next year as certified seed from which 10-15 acres commercial cultivation can be easily done. Nowadays single eye bud/ bud chip technology with polytrays is being rapidly adopted by the sugarcane growers. If TCP's as breeder seed for planting mother seed plot and later stages with single eye bud is followed the coverage of area under quality seed will be three to four times higher at each stage. Further the cost of quality seed production will drastically be reduced. Seed production, seed distribution and other related aspects, therefore need to be improved and strengthened at the farmers' level which will increase the Seed Replacement Rate (SRR). Despite the implementation of the organized seed programme in sugarcane, the seed replacement rate has only reached the level of 15%, and 85% of the seed used is farm saved during the harvesting cane. There exists an alarming gap between the demand and supply of quality seeds with a vast scope to produce and distribute quality seeds in this crop. For this purpose, seed village concept is a novel and highly practical approach to facilitate production and timely distribution of quality seeds of desired varieties. The seed village concept also advocates rural self employment as well as self-sufficiency in production and distribution of quality seeds.

STATUS OF ADOPTION

As compared to conventional seed, the use of TCP's appears to be a costly affair because of high cost of production. The present production cost at VSI is Rs. 7.50 per plantlet. The Institute provides the plantlets at the subsidized rate of Rs.6.00/- per plantlet to sugar factories and sugarcane growers in Maharashtra. During the past 15 years, VSI has been striving for the popularization of this technology among the sugarcane growers. There has been an increased awareness among the sugar factories and farmers about the benefits of this technology and the demand for tissue cultured planting material is increasing day by day. However the progress is very slow.

Table 2. Supply of tissue cultured plantlets to sugar industry

Year	Production	Distribution	Research Use
1999-2000	2,81,627	98,245	40,063
2000-2001	5,29,599	3,67,489	82,582
2001-2002	16,21,216	8,44,835	24,900
2002-2003	24,24,441	11,81,681	1,65,514
2003-2004	13,84,208	10,74,058	1,79,760
2004-2005	19,80,274	10,93,311	35,288
2005-2006	19,21,050	13,86,980	24,180
2006-2007	21,06,120	15,19,755	74,525
2007 -2008	18,00,000	14,06,503	34,794
2008-2009	5,42,937	4560000	31,850
2009-2010	12,25,150	1040876	2765
2010-2011	12,02,500	1038982	5500
2011-2012	1052820	907134	-
2012-2013	10,52,000	9,95,640	9,737
Total	19123942	17515468	218639

The status of TCP's utilization by sugar factories/farmers in Maharashtra and adjoining states is given in Table 2. In Maharashtra, average area under sugarcane cultivation is around

8 lakh hectares. About 40 % area comes under ratoons crop. For the remaining area under three tier seed multiplication program, about 150-160 ha breeder seed plots through tissue cultured plantlets will be required, i.e. 25-30 lakh TCP's has to be raised. Maharashtra's average consumption of TCP's in the last ten years is about 8-10 lakh. On this basis it can be said that there is only 25-30 % adoption for the TCP's. However, from Table 2 further it can be seen that there is no stable demand for TCP's. This is because the fluctuations in environmental conditions and pricing for sugarcane. Whenever there is a favorable climate and good price for sugarcane, demand for TCP has increased. This has imposed difficulty for the tissue culture laboratories to utilize their full capacity of production and hence, under low capacity utilization, the cost of production will increase.

Table 3. State wise Supply of plantlets by VSI

State	Plantlets Supplied
Maharashtra	1,05,87,899
Gujarat	2538500
Madhya Pradesh	782485
Uttar Pradesh	49300
Karnataka	63860
Goa	98343
Andhra Pradesh	137360
Rajasthan	12200
Bihar	26100

STRATEGY TO INCREASING ADAPTABILITY OF TCP'S

Since sugarcane is a vegetatively propagated crop, many farmers use commercial cane from their own fields as seed cane which may carry pest and diseases. Repeated use of the commercial cane for planting is one of the reasons for low productivity. Therefore, proper strategy should be implemented to promote the adoption of tissue culture technology in sugarcane to increase sugarcane and sugar yield. In this regard, following suggestions may be helpful for developing a policy to strengthen quality sugarcane seed production at grass root level of sugarcane growers.

a) Efficient Management of Sugarcane Micropropagation in Tissue Culture Laboratories

Sugarcane micropropagation is not widely accepted by commercial tissue culture laboratories. If demand for sugarcane TCP's for the country is considered, there is ample scope for these laboratories as sugar industry is growing very rapidly and there is great demand for quality seed for better cane productivity as well as supply of cane for crushing. The planting density in sugarcane is much higher than the horticulture crops, and hence expecting huge demand for TCP's, several laboratories have initiated sugarcane micropropagation. Later on, these laboratories diverted their priority to other commercial/horticulture crops as it e was not much viable only with sugarcane. The reason for this may be the cost of production of TCP's. It is therefore essential to reduce the production cost of TCP's in sugarcane. The laboratories producing minimum 5 lakh plantlets/annum are becoming economically viable with strict control over different parameters such as-

- Capacity utilization should be more than 80%.
- Culture rejection and contamination rate must be below 5%.
- Mortality in green house should not be more than 5%.
- Appropriate varietal balance should be followed in such a way that capacity utilization should be as much uniform as possible.
- Apart from high and medium demand varieties, low demand varieties should also be taken up.
- Low cost technologies reported for tissue culture viz. use of commercial sugar, replacement of agar, use of natural light and LED lights, solar water heaters should be incorporated in the micropropagation technology [1, 4, 19, 20 and 25].
- The continuity of Scientific and Managerial manpower should be ensured.
- The management information system (MIS) for production planning and work scheduling should be properly utilized to minimize the operational cost.
- The manpower use (operators/labors), electricity and water resources should be efficiently utilized for minimizing recurring expenditure.

The labor wages and electricity are the main cost contributing factors. The use of water purification systems, media dispensing unit, bottle washing machines in the laboratory, soil mixing and filling of poly-bags through machines, sprinkler irrigation system for TCP irrigation and sprays of pesticides/ nutrients can reduce the recurring expenditure considerably. With several innovations and modifications in media components and protocols, the production cost can be reduced considerably [6, 7]. The use of natural resources/innovations for lighting may be utilized to reduce the electricity consumption. The tissue culture laboratories therefore need to be designed so as to make the production of TCPs economically viable.

b) Field Demonstrations to the Farmers

The importance of superior planting material/technologies over the conventional systems should be demonstrated on farmer's field to increase the adoption rate. The initial cost of TCP's is high due to which farmers do not prefer to them in three tier seed production program. For calculating the cost, all the three stages should be considered together instead of TCP's cost only. In comparison to conventional seed production system, TCP's use is considered economical because of the cost of production at three stages of seed production, the seed material obtained for commercial cultivation and benefits due to quality seed.

c) Convincing for Farmers' Over Expectations

Farmers consider TCP's as magical entities and expect better output even under traditional cultivation practices. It is therefore necessary that farmers should be convinced about the technology and that the outcome is achievable only if proper agriculture practices recommended by the R&D Institutes/Universities are followed at the right time. The packages of agronomical practices therefore need to be demonstrated at farmer's field to disseminate the right technology. If the comparative performance of crops raised under recommended and traditional practices is clear, it ensures the right message of the TCP's to farmers.

d) Strengthening of Extension Services

During the distribution of TCP's, the information leaflet detailed with all the agronomical practices should be provided. The seed nurseries should be regularly visited for timely guidance to the farmers. Sugar mills should come forward to implement the seed village concept at block level in their area of operation. This will reduce the expenditure on seed transportation from seed plots at research stations/ universities to farmer's field and farmers will get fresh quality seed nearby their plantation site. This will minimize delay in harvesting to planting and ensure proper management of resources. The state government's extension staff at block levels that play an important role in technology transfer should be thoroughly trained for the implementation of technology. The follow up on seed plots should be regularly done so that the quality seed plots will be raised and canes will be strictly harvested as seed cane and not for crushing purpose. Demand and supply in adjoining areas and outside factory operation area need to be interlinked the throughout the sugarcane growing regions so that quality cane could be utilized only for seed purposes. This will also prevent the farmer for supplying the canes for crushing, if there is no demand for seed in his village. There should be proper advertising for seed and finance availability as well as the benefits. For this purpose, the local news papers /electronic media like television / SMS services needs to be exploited.

e) Care of Seed Nurseries

It is commonly observed that even though information pamphlets are provided along with the plantlets, traditional agronomical practices are still followed resulting in heavy tillering and thin canes. Ignoring the recommended agronomical practices can result in crop failure, leaving a bad impression with the farmers about the technology. Sugarcane growers should be properly trained about the care of plantlets before planting, at planting and, after planting for better survival/germination, proper tillering, quality cane formation with higher number of internodes/cane and canes per stool.

f) Imbalance in Production Capacity and Consumption Area

There is no pre-planned demand for TCP's. The farmers demand for the plantlets sometimes can come without advance intimation which can makes the tissue culture laboratories difficult to fulfill the demand and in some cases plants are supplied without sufficient hardening cause higher mortality in the field. If the demand is placed well in advance then tissue culture laboratories can plan production and the supply in time. Further, due to non-availability of plantlets in nearby areas, TCP's are brought from distant places which again increase the transportation cost and mortality. If seed village concept is implemented properly, the availability of quality seed can be ensured at the door steps of farms with appropriate time, with affordable cost. There will be increased confidence among the farmers about the quality because of the known source of production. It will also be helpful for the TCP's producing laboratories/ seed farms to manage their full capacity utilization.

g) Availability of Micro-Credit

The cost of tissue-cultured plantlets is comparatively high. The TCP's are produced by research centers or universities and the transportation cost to fetch the plantlets at plantation site is high, often more than the cost of plantlets. Thus small farmers with limited resources find it unaffordable for the quality seed and they continue with the traditional material. Therefore, the micro-credit for the purchase of tissue culture plantlets should be provided. If sugar mills club the demand and arrange transport facility then transportation cost will be reduced considerably and sugarcane growers will get quality seed for plantation of seed nurseries. To upgrade the quality of farmer-saved seed (which is about 80-85% of the total seed used for crop production program), it is proposed to provide more financial assistance for distribution of foundation/certified seed. Fortunately, subsidy through the State Government is provided to sugarcane growers through Mega Seed Project sponsored by the Central Government; but many of the sugarcane growers and even some extension officers in the Govt. Agriculture Department are unaware about this provision. In order to strengthen seed village concept farmers should be encouraged to participate in raising seed plots of appropriate quality.

CONCLUSION

There is no adequate availability of quality seed-cane with sugar mills and therefore the rate of seed replacement is not as per requirement. There is a huge gap between demand and supply of quality seed. Awareness in application of tissue culture technique in seed rearing at common farmer level is either lacking or absent. Sugarcane micropropagation (TCP's) therefore, needs to be encouraged for assured genetic purity, better and early germination, quicker coverage by superior varieties and high sugar as well as sugarcane yield. Adoption of three-tier seed nursery programme with seed replacement at least once in 4 years can be implemented. Increasing use of tissue culture seed for multiplying the basic genetic material needs to be supported to develop 50% seed replacement through tissue culture technique and 50% through conventional seed program. Thus successful use of TCP's requires appropriate integration with other sources of science-based and traditional knowledge for sustainable agriculture development in the country.

REFERENCES

[1] Dalvi S. G., Gudhate P. G. and Theertha Prasad D. (2011) Low cost support matrix for Potato Micropropagation. *Potato Journal,* Vol. 38, No.1-2, 2011.

[2] Dalvi S.G. (2007). "Production and certification of tissue cultured plantlets" National Level Seminar on Sugarcane Seed Production and Certification, at VSI Pune, 22-23 Feb. 2007 pp. 61-71.

[3] Dalvi S. G. and Tawar P. N. (2006). Micropropagation technology for quality seed in sugarcane. Proceeding of State level Cane Development Workshop "Recent developments in Sugarcane Production Technologies with reference to precision farming" for Managing Directors, Cane Development Officers from Sugar factories held at VSI on 13th & 14rh Sept.2006 pp G1-6.

[4] Dalvi S. G., Kamat V. A., Dixit G. B., Prasad T. D., Tawar P.N. and Deshmukh R.B. (2011) Absorbent cotton as support matrix improves morpho-physiological responses in sugarcane callus by buffering medium pH and increased nutrient uptake. Presented in 4th IAPSIT International Sugar Conference and Expo, (IS-2011) New Delhi, 21-25 November, 2011. pp 76-766.

[5] Devarumath R. M., Doule R. B., Kawar P. G., Naikebawane S.B. and Nerker Y.S. (2007) Field performance and RAPD analysis to evaluate genetic fidelity of tissue culture raised plants vis-a-vis conventional setts derived plants of sugarcane. *Sugar Tech* 9(1): 17-22.

[6] Geetha S. and Padmanabhan D. (2002) Evaluation of Tissue Culture Raised Sugarcane for Yield and Quality. *Sugar Tech,* Vol. 4 (3&4): 179–180.

[7] Hendre R. R., Mascarenhas A. F., Nadgir A. L., Pathak M. and V. Jagannathan (1975). Growth of sugarcane mosaic virus free sugarcane plants from apical meristems. *Indian Phytopathology* 28:175–178.

[8] Iyyar R. S., Hendre R. R., Kotwal M., Khuspe S. S. and Mascarenhas A. F. (1983). Rapid multiplication of sugarcane through tissue culture. *Sugarcane* 1: 5–7.

[9] Jalaja, N. C. (1994). Sugarcane Adaptive Research Project. Mid-term appraisal report, 1989-1994. Micropropagation. Sugarcane Breeding Institute, Coimbatore. pp. 22-31.

[10] Jalaja, N. C. (2001). Micropropagation of sugarcane varieties for quality seed production. In: Manual on Sugarcane Production Technology. Sugarcane Breeding Institute, Coimbatore. Pp 48-52.

[11] Nerkar Y. S., Pant N. M., Tawar P. N., Sawant R. A., Dalvi S. G., Nikam A. A., and Devrumath R. M. (2006). Protocol for production and supply of quality seed in sugarcane using micropropagation technology" Lead paper on "Developing standards for tissue culture raised planting material of Sugarcane" *Proceeding of Brain storming session* organized by ICAR & IISR Lucknow at VSI dated 9th June 2006 pp. 5-15.

[12] Lal J. and Pande H. P. (2003). *In vitro* rejuvenation and micropropagation of deteriorated sugarcane cultivars. *Indian Journal of Sugarcane Technology,* 18 (l&2): 85 87.

[13] Lal, N. (1997). Performance of sugarcane mericlone transplants under different inter and intra-row spacing combinations. *Sugarcane* 6:12-14.

[14] Lal, N. and Krishna R. (1997). Yield comparison in sugarcane crop raised from conventional and mericlone derived seed cane. *Indian Sugar* XLVII (8): 617-621.

[15] Lal M., Singh R. K. Srivastava S. Singh N. Singh S. P. and Sharma M. L. (2008). RAPD marker based analysis of micropropagated plantlets of sugarcane for early evaluation of genetic fidelity *Sugar Tech* 10(1): 99-103.

[16] Pawar S.V., Patil, S. C., Jambhale V. M., Naik R.M. and Mehetre S. S. (2002). Rapid multiplication of commercial sugarcane varieties through tissue culture. *Indian Sugar,* 52(3) 183-186.

[17] Ramanand, Lal M. and Singh S. B. (2005) Comparative Performance of Micropropagated and Conventionally Raised Crops of Sugarcane. Sugar Tech 7: 93-95.

[18] Rani V. and Raina S. N. (2000). Genetic fidelity of organized meristem-derived micropropagated plants. A critical reappraisal. *In Vitro Cell Dev. Biol. - Plant.* 36**:** 319-330.

[19] Savangikar, V. A. (1995). Sugarcane tissue culture technology development and commercial side. Seminar on application of Biotechnology for sustainable sugarcane production. 44 ~ Annual Conv. DSTA, Pune, pp. 48-51.

[20] Savangikar, V. A. (1996). Observation on the performance of commercial tissue cultured plantation at low planting density.44 ~ Annual Conv. DSTA, Pune, pp. 144-148.

[21] Singh, B., Yadav, G. C. and Lal M. (2001). An efficient protocol for micropropagation of sugarcane using shoot tip explants. *Sugar Tech.,* 3(3): 113-116.

[22] Sood, N., Srivastava, R. and Gosal S. S. (1999). *In vitro* multiplication and field evaluation of sugarcane. *Proc. STAI,* 61: 3-9.-186.

[23] Sreenivasan, T. V and Jalaja N. C. (1992). Micropropagation of sugarcane varieties for increasing cane yield. *SISSTA Jour.* 19**:** 61-64.

[24] Tawar P.N. Sawant R. A., Dalvi S. G., Nikam A. A., Kawar P. G., and Devarumath R.M. An assessment of somaclonal variation in micropropagated plants of sugarcane by RAPD markers. *Sugar Tech* 2008, 10(2)124-127.

[25] Yadav S., Saini N.and Jain R. K. (2004) Low cost multiplication and RAPD analysis of micropropagated plants in sugarcane. *Physiol Mol Boil Plants* 10(2):269-276.

[26] http://www.apcoab.org/uploads/files/1279706285sugar_pub.pdf

[27] http://www.dbtncstcp.nic.in/downloads/Sugarecane.pdf

[28] http://www.iisr.nic.in/download/publications/RoleofMicropropagation-enlish.pdf

In: Current Status of Sugarcane Research in India
Editors: A. K. Tiwari, M. Lal and A. K. Singh
ISBN: 978-1-63463-458-8

Chapter 6

FERMENTATIVE PRODUCTION OF SUGARCANE VINEGAR

G. S. Kocher **[*]** ***and H. K. Dhillon***
Department of Microbiology, Punjab Agricultural University, Ludhiana, India

ABSTRACT

Fermented vinegar is a known food, tonic and medicine. In spite of its properties, people in India primarily consume synthetic vinegar mainly because of the lack of awareness and comparatively higher cost in respect of natural vinegar. Therefore, a cost effective and microbiologically safe fermentation process for sugarcane vinegar was optimized at 50 L batch and 5 L semicontinuous scales using indigenous low cost fermenters. The latter was a fibreglass fermenter which produced a quality vinegar (in terms of percent acidity, number of days and organoleptic quality) at par with stainless steel fermenter under batch fermentation conditions. Similarly, a PVC fermenter containing wood shaving's adsorbed cells of *Acetobacter aceti* was optimized to produce vinegar in 3 days under semicontinuous fermentation conditions.

Keywords: Acetic acid fermentation, Amino acids, Fermentation batches, Minerals, Vinegar, Sugarcane juice

INTRODUCTION

Vinegar is a household food supplement which is known to improve digestion, add flavour and also act as an excellent food preservative. It is basically of two types, synthetic (prepared from glacial acetic acid) and natural (produced by fermentation of sugar rich fruits) vinegar. In India, synthetic vinegar, which is a 4 % solution of acetic acid in water blended with caramel and essences is commonly used on commercial scale in spite of the fact that it has practically no food value [1]. Natural vinegar, on the other hand involves two successive

---

[*] E-mail: gskocher@pau.edu.

biochemical processes; a primary yeastous fermentation of sugars to ethanol followed by secondary fermentation of ethanol by acetic acid bacteria to acetic acid [2, 3]. Besides, there are commendable health benefits of natural/fermented vinegar as its gastronomic value has been appreciated for thousands of years. Along with this, natural vinegar possesses antibacterialy activity, lowers high blood pressure, reduces total cholesterol and thus lowers risk of cardiovascular disease [4, 5] and also has antioxidant activity. On the other hand synthetic vinegar may cause poisoning, gastro enteric illness and diarrhoea besides developing brain fever in young children [6]. Hence, there is a need to generate consumer awareness in respect of the health benefits of natural vinegar and promote its production at farmer/ small entrepreneur level.

The growing social and economic significance of fermented vinegar has fostered research into the most salient aspects of its production processes [7]. Natural vinegar may be prepared from any sugary material such as molasses, fruits, honey, cereals, potato, whey, sugarcane etc. Among the different raw materials available for vinegar production, sugarcane is an optimal choice due to its high sugar content and easy availability. Sugarcane cultivation occupies a prominent position on the agricultural map of India covering large areas in sub-tropics and tropics. It is the sole raw material for the largest agro-processing industry in the rural sector, wherein about 6 million growers cultivate this crop. Sugarcane is a multi-product crop and has immense potential for diversification. Besides the production of sugar, green top of sugarcane is used as fodder for milk cattle and molasses, a by-product of sugar processing is used as cattle feed.

Although a wealth of knowledge currently exists on vinegar production process, the very small industry dealing with the production of natural vinegar still employs the traditional batch fermentation using free acetic acid bacterial cells which generally spans 4-5 weeks [8, 9, 10]. The immobilized cell technology, on the other hand is advantageous over free cells employed in batch fermentation as it improves upon the fermentation process by increasing biomass, option of reusability, protection of cells from toxic effects of low pH, temperature, inhibitors etc. besides keeping the conditions aerobic which are necessary for acetic acid bacteria [11, 12, 13]. Further, cell immobilization helps in early clarification of the product [14]. Among the different immobilization techniques, adsorption reduces the problems associated with oxygen diffusion and does not suffer from the scale up drawbacks [15, 13]. Further, the choice of immobilization material in the form of inexpensive and easily available inert biological materials can help reduce the cost of the process. In this regard a number of supporting materials and carriers such as hollow fibres [16], cotton towel cloth [13], polyurethane [17], wood shavings, corn cobs and baggase [14] have been used for immobilization of *Acetobacter* cells.

It is thus imperative to develop a cost effective process which reduces fermentation time as well as increases the concentration of acetic acid in the vinegar. Hence, production of natural vinegar was studied under batch and semi continuous conditions. In the present study, a 50L batch and a 5L semicontinuous fermentation technologies were optimized using low cost fermenters as part of a DST sponsored technology demonstration project. This technology will enable entrepreneurs to produce natural sugarcane vinegar and earn revenue out of it. Also, a common man or a hobbyist will be able to produce natural vinegar at affordable costs and thus will not have to depend on the synthetic vinegar available in the market.

PREPARATION OF FERMENTED SUGARCANE VINEGAR

Processing of Canes

As sugarcane was substrate of choice, the foremost step involved selection of canes by first eliminating all the damaged and spoiled ones. It was followed by washing them under running water in order to remove dirt and adhered microorganisms. Canes were later crushed in a mechanical crusher and the juice was heated in an aluminium tub at 75-85°C to pasteurize and remove haze (due to dextran) from it. The juice was siphoned thereafter using a sterilizable PEP pipe of diameter 5mm. Earlier, Pretreatment of sugarcane juice has been reported by pasteurization at 85°C for 15 min [18].

Ethanolic Fermentation

Vinegar production is a two step fermentation procedure. The first step is the alcoholic fermentation of juice to ethanol which is followed by secondary fermentation of ethanol to acetic acid by acetic acid fermentation. Two types of fermenters, one made from stainless steel (SS) and second of fibre glass (FG) of capacity 75 L were designed to carry out fermentation at 50L scale (Figure 1). The SS fermenter in which alcoholic fermentation was carried out was equipped with a stainless steel (SS118) cylindrical vessel of 75 L capacity with a height-to-diameter ratio of 2:1. The aeration in the SS fermenter was controlled with an air supply system consisting of air filters and inlet pipe with sparger; a chiller to maintain temperature and prevent loss of volatile components; an electrically heated jacket and a sensor for temperature measurement. On the other hand, the fibre glass fermenter was supplied with sufficient air using a submerged submersible pump (NOVA) and temperature was controlled by keeping it in a temperature regulated room. The 50 L fermentation medium was inoculated (@ 5% v/v) with an overnight grown culture of *S. cerevisiae* MTCC 11815 (prepared in sterile 15% jaggery solution) at 28±2°C. Samples were withdrawn every 24 h till completion of fermentation. The cell free supernatants were stored in a deep freezer and later analysed for TSS (using a Digital Brixometer (ATAGO, Japan) and glass brixometer), ethanol [28] and pH (pH meter, Hanna Instruments, China).

The fermentation of the above juice (17°Bx) by *S. cerevisiae* presented a complete sugar utilization in 3-4 days with an ethanol production of 9.9 to 10% (v/v) and corresponding fermentation efficiencies of 91.5 to 92.5%, respectively in FG and SS fermenters (Table 1). Several workers have reported incomplete to complete utilization of sugars during alcoholic fermentation. Earlier studies in our laboratory on alcoholic fermentation of sugarcane juice (17°B) also revealed complete utilization of sugars [46]. In the studies carried out by Lashley [18] for the production of alcoholic beverages from sugarcane juice, an ethanol content of 11-12% by weight has been reported. Kida et al. [19] reported that repeated batch fermentation of molasses medium containing 20% (w/v) sugars at 30°C produced 8.5% (w/v) ethanol with efficiency of 83.3% whereas Kaur et al. [33] reported 8.98% alcohol production from molasses with a fermentation efficiency of 88%.

The data also exhibited a direct correlation between ethanol production and decrease in pH with a drop in its value from 5.5 to 3.2 (Table 1) which can be attributed to release of organic acids and CO_2 during fermentation [20, 21, 22, 23].

The pasteurized sugarcane juice had 97.8mg/100ml of ascorbic acid while it was 88.8mg/100ml in the sugarcane alcohol, thus retaining a large part of ascorbic acid after fermentation. Such retention has also been reported earlier by Muhammad et al. [24]. An increase in the total phenols was noticed in the sugarcane alcohol (142.5mg/100ml) as compared to the juice (56.1mg/100ml, Table 1). This is attributed to the fact that fermentation increases extraction of phenols [25] and some phenols are also formed during fermentation [26]. Thereafter, the produced alcoholic wort was kept undisturbed in a cool place for 2 days and decanted for carrying out its secondary fermentation by acetic acid bacteria.

Acetic Acid Fermentation

Batch and Semicontinuous Acetic Acid Fermentation

The preparation of inoculum in respect of acetic acid bacteria (7.5%, v/v) takes 3-4 days, hence this process was initiated on the third day of the alcoholic fermentation so that it is available by the time the decanted alcoholic wort is ready. The bacterial inoculum (seed culture) was prepared in 250 ml sterile 1.5% jaggery solution. This seed inoculum was used to inoculate a 3.75L (1.5%) jaggery solution and incubated at 28-30°C for 36-48 h. The semicontinuous vinegar fermentation was carried out with immobilized cells of *A. aceti* (an isolate from sapota fruit) in an indigenously designed PVC column having a height: diameter of 45: 15 cm (Figure 1). Shredded, dried and steam sterilized carrier material (wood shavings, 2.5 Kg) was packed, covering about 1/3rd of the column. The latter was drenched with sterilized water to its maximum water holding capacity till water started oozing out of the carrier materials in order to remove any color of the sterilized and packed wood shavings. Thereafter, 36h old *A.aceti* broth culture was percolated through the column maintaining cells : adsorbent ratio of 2:1, and left to stand for about 12 h for adsorption of cells by adhesion forces between two solid surfaces i.e. carrier material and cells [27]. The immobilized cells were then subjected to an overnight adaptation by adding a mixture of 7.5% (v/v) ethanol fermented sugarcane ethanol and 10% (v/v) mother vinegar to the packed column. Sugarcane ethanol (supplemented with mother vinegar) with an initial acidity of 2% was trickled @ 50ml/h through the packed bed at 5.0 L scale while temperature was maintained at 28±2°C. Simultaneously, a batch fermentation was carried out in a sterile plastic can of 10L capacity at 5L scale. The decanted alcoholic wort was mixed with 10% (v/v) mother vinegar and 7.5% (v/v) previously prepared bacterial inoculum. The alcoholic wort was also supplemented with a nutrient solution consisting of 0.1% (v/v) each of $(NH_4)_2SO_4$ and KH_2PO_4 and 1g/100 ml of stock solution consisting of yeast extract, sucrose, $ZnSO_4$, $MgSO_4$ and $MnSO_4$. The inoculated wort was incubated at 28-30°C, with continuous mixing for 3 days using a submersible pump. Thereafter, the fermentation can was left stationary. When the bacterial pellicle started breaking or settling down, the supernatant was collected and allowed to settle for a day before it was filled in bottles. The latter were pasteurized in a water bath at 80°C for 5 min and stored at ambient temperature.

The periodic samples collected aspetically both from batch and semicontinuous (after each cycle of 50h) fermenters were analysed for ethanol content [28], volatile acidity [29], total phenols [30], ascorbic acid [31] and pH.

Figure 1. Indigenous Fermenters (Stainless steel, A; Fibre Glass, B and PVC, C).

Table 1. Fermentation of sugarcane juice by *Saccharomyces cerevisiae* MTCC 11815 for ethanol production

Type of fermenter	Alcoholic Fermentation					
	Initial Brix (°B)	Ethanol Produced (% v/v)	Number of days	pH		Fermentation Efficiency (%)
				Initial	Final	
Stainless steel (SS)	17.00±1.7	10.0±0.68	4.0±1.0	5.5±0.1	3.5±0.3	92.55±2.32
Fibre glass (FG)	17.00±2.6	9.9±0.18	4.0±1.0	5.3±0.1	3.2±0.3	91.57±1.70
Total Phenols (mg/100ml)	142.5±0.20; (56.1±0.18)*					
Ascorbic acid (mg/100ml)	87.8±0.24; (97.8±0.21)*					

The values are means of triplicate experiments.
*Values in the brackets indicate the respective values in juice.

The semicontinuous vinegar fermentation using adsorbed cells of *Acetobacter aceti* AC1 was found to be complete in 3 days, consistently for 25 cycles except the first cycle, which showed a lag phase and hence took 7 days to produce the required volatile acidity of 4% (w/v) (Figure 2). Before commencing each cycle, the initial acidity was kept 2% because inhibition of acetic acid fermentation above 2% has been reported earlier by Nanba et al. [16]. The volatile acidity in respect of 25 cycles was found to vary between 4.12 to 6.84% (w/v). Earlier, similar acidity values have been reported by different workers under fed batch fermentation conditions [32, 17]. The semicontinuous vinegar fermentation was approximately 4 times faster than the batch process which involved 13 days of fermentation to produce vinegar with a volatile acidity of 5.80% (Figure 2, Table 2). The fermentation efficiency was found to be variable in different fermentation cycles with highest efficiency of 87.9%, which is much more than that reported by Kaur et al. [33] in tea vinegar production using immobilized cells in trickling fermentation (42.7%). In batch fermentation, the fermentation efficiency was 74.5%. Earlier, aeration has been considered an important parameter for acetic acid fermentation as it helps in increasing biomass that enhances the rate of acetic acid fermentation [34, 35, 36]. The partial aeration by agitation reported by us has not been described earlier to the best of our knowledge as literature reports continuous shaking [32, 17]. Hence, our technique of employing partial aeration during acetic acid fermentation is a less energy consuming process. Also, the use of multiple nutrients like ammonium sulphate, yeast extract, cornsteep liquor, bionyl F15, ammonia and potassium dihydrogen phosphate, sodium hydrogen phosphate as source of nitrogen and phosphorous, respectively [32, 37, 38, 39] for obtaining high acetic acid fermentation rates has been reported earlier [40, 38, 39, 36].

The semicontinuous vinegar fermentation was more efficient, as interpreted from acetification rates that were 22.8/L/day and 4.46g/L/day under semicontinuous and batch fermentation cycles, respectively. According to Mehaia and Cheryan [41], vinegar productivity using membrane bioreactor was 20 times more than free cells whereas Ory et al. [17] reported a vinegar productivity of 4.74 g/L/ day using polyurethane adsorbed cells of *Acetobacter aceti* at 42.7% of fermentation efficiency. Garg et al. [42] in their studies on semicontinuous production of mango vinegar by *A.aceti* cells immobilized on wood shavings reported a final acidity of 5.3% (w/v) with 60% vinegar recovery. The adsorbed cells showed better fermentation efficiency because of the increased surface area and direct contact between the substrate and air spaces within the adsorbed materials [17]. Earlier in our laboratory, sugarcane vinegar was produced using sugarcane bagasse, corn cobs, wood shavings and calcium alginate immobilized cells in 8 rounds (10 h, each) at 0.5 L fermentation scale which revealed wood shavings as better than the other immobilization materials studied [14]. Ory et al. [17] their studies reported polyurethane foam to be a better adsorbent of acetic acid bacteria.

Total phenols at the end of different semicontinuous fermentation cycles ranged from 81.9±0.14 to 138.6±0.49 g/ml which was similar to the value of phenols found in vinegar produced through batch method (110.0 mg/100ml, Table 2). Earlier workers have also reported the presence of different types of phenolics in fermented vinegar using HPLC technique but we did not come across any report on determination of phenols in sugarcane vinegar. Elsewhere, Mraques et al. [43] reported 43.27mg/100ml of total phenols and polyphenols in honey blended orange vinegar. Till date, no work has been carried out regarding the ascorbic acid content in the sugarcane vinegars though references [44, 45] on sugarcane juice (470mg/100ml) are available in literature. In the present study, 42.0±1.13 to 86.1±0.77 mg of ascorbic acid/100mg of sample was detected which was again consistent with the ascorbic acid content of the vinegar produced through batch fermentation (72 mg/100ml). As ascorbic acid is heat sensitive, pasteurization and sterilization had resulted in its reduction compared to its content in juice though fermentation led to retention of ascorbic acid in the sugarcane vinegar. Decreased levels of ascorbic acid were also reported by Muhammad et al. [24] during pasteurization of strawberry juice.

Table 2. Characterization of Sugarcane vinegar produced by immobilized cells in different cycles of semicontinuous fermentation

Lot No	% Volatile Acidity (v/v)	% Fermentation efficiency	Yield Factor	Acetification rate (g/l/day)	Total phenols (mg/100ml)	Ascorbic acid (mg /100ml)
1	4.1±0.04	52.6±0.12	0.54±0.07	5.85±0.07	81.9±0.21	59.8±2.30
2	4.4±0.01	56.5±0.31	0.58±0.31	14.6±0.31	81.9±0.14	86.5±1.34
3	4.12±0.04	52.9±0.07	0.54±0.12	13.7±0.12	110.5±0.70	71.2±0.28
4	4.8±0.03	61.6±0.12	0.64±0.07	16.0±0.03	115.6±0.49	74.7±4.66
5	6.39±0.03	87.3±0.12	0.85±0.02	21.3±0.06	115.7±1.06	45.4±3.67
6	4.2±0.02	53.9±0.32	0.56±0.31	14.0±0.10	133.2±0.28	84.8±1.13
7	4.2±0.04	53.9±0.31	0.56±0.11	13.2±0.31	131.7±0.42	60.7±1.06
8	6.8±0.07	87.3±0.07	0.90±0.17	22.6±0.09	118.5±0.63	71.7±0.42
9	5.88±0.05	75.5±0.17	0.78±0.07	19.6±0.17	117.5±0.77	86.1±0.77
10	4.52±0.1	58.0±0.12	0.60±0.14	15.0±0.04	99.7±0.35	85.1±0.70
11	4.8±0.02	61.6±0.11	0.64±0.15	16.0±0.21	83.5±0.70	72.2±1.13
12	4.42±0.30	56.5±0.31	0.58±0.11	14.7±0.30	90.4±0.77	61.7±0.35
13	5.1±0.22	65.5±0.19	0.68±0.31	17.0±0.31	110.9±0.14	60.2±1.76
14	5.4±0.21	69.3±0.07	0.72±0.32	18.0±0.32	101.4±0.84	84.3±1.83
15	4.68±0.12	60.1±0.12	0.62±0.31	15.6±0.23	138.6±0.49	45.6±4.03
16	5.04±0.19	64.7±0.03	0.67±0.32	16.8±0.21	117.6±0.49	43.5±1.06
17	4.08±0.02	52.4±0.42	0.54±0.25	13.6±0.24	99.9±0.14	42.0±1.13
18	4.36±0.01	57.3±0.31	0.58±0.07	14.5±0.31	83.7±1.06	71.7±0.42
19	6.28±0.21	80.7±0.23	0.83±0.22	20.9±0.24	88.4±1.97	71.2±0.28
20	6.72±0.22	86.3±0.07	0.89±0.23	22.4±0.22	88.9±0.14	84.8±1.13
21	5.76±0.05	74.0±0.15	0.76±0.31	19.2±0.07	111.1±1.2	60.7±1.06
22	4.98±0.05	64.0±0.07	0.66±0.21	16.6±0.25	95.7±0.42	61.3±0.21
23	6.0±0.04	77.1±0.11	0.80±0.01	20.0±0.21	98.4±2.05	84.8±1.13
24	6.84±0.31	87.9±0.12	0.91±0.23	22.8±0.12	118.0±0.07	60.7±1.06
25	6.84±0.01	87.9±0.31	0.91±0.17	22.8±0.31	117.5±0.77	71.7±0.42
Batch fermentation	5.80±0.10	74.5±0.10	0.77±0.21	4.46±0.37	80.8±0.17	75.8±0.80

*Fermentation conditions:

Temperature: 28±2°C; Initial alcohol: 7.5% (v/v); Mother Vinegar: 10% (v/v).

Initial acidity: 2% (v/v).

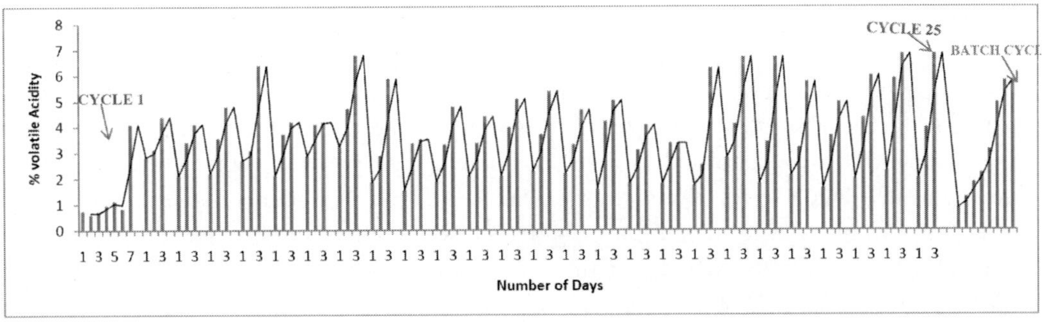

Figure 2. Comparative production of acetic acid by immobilized cell technique (semicontinuous process) and batch process.

Storage and Quality Analysis

The quality analysis of sugarcane vinegar was evaluated with respect to its sensory score (on the basis of modified Hedonic scale by 10 semi trained judges with scoring points as: 9-10, Outstanding vinegar; 7-8.99, Standard vinegar; 5-6.99, Commercial vinegar; 3-4.99,

Below commercial vinegar acceptability and 1-2.99, Spoiled vinegar [46], shelf life (on the basis of volatile acidity [29], total phenols [30], ascorbic acid [31] amino acids (qualitatively by thin layer chromatography using Silica gel G as adsorbent and mobile phase consisting of Butanol, acetic acid and water in the ratio of 40:10:30) and minerals by Atomic absorption spectrometry (through outsourcing, Department of Soils, Punjab Agricultural University, Ludhiana).

Table 3. Sensory evaluation of vinegar produced by semicontinuous and batch scale fermentation

Sensory Analysis	Maximum Points	Vinegar fermentation	
		Semicontinuous	Batch
Appearance	2	1.68±0.08	1.66±0.01
Colour	2	1.70±0.05	1.53±0.05
Astringency	2	1.74±0.04	1.67±0.06
Sourness	2	1.78±0.06	1.57±0.01
Bouquet	2	1.76±0.10	1.56±0.02
Total	10	8.66±0.33	7.99±0.15

The above scoring is mean (±) standard deviation of evaluation by 10 panelists.

Sensory Quality

9-10: Outstanding vinegar
7-8.99: Standard vinegar
5-6.99: Commercial vinegar
3-4.99: Below commercial vinegar acceptability
1-2.99: Spoiled vinegar

The vinegar produced by semicontinuous fermentation in 25 cycles was also found to be consistent in terms of sensory attributes. It was found to be acceptable with a mean sensory score of (8.66±0.33) on modified hedonic scale, whereas mean sensory score of vinegar produced through batch fermentation was 7.99±0.15 (Table 3). The results revealed that the vinegar produced by both fermentation methods was of standard quality, revealing that there is not much difference in the sensory qualities of vinegar produced through either method. In literature, Tesfaye et al. [47] studied the sensory characteristics of sherry wine vinegar aged in oak casks.

The TLC analysis revealed that the vinegars were enriched in seven amino acids (alanine, isoleucine, valine, leucine, valine, tryptophan, methionine and histidine) of which except alanine, the other six were essential amino acids (Figure 3). The atomic absorption spectral mineral analysis carried out by outsourcing revealed five major minerals i.e. Potassium, Magnesium, Sodium, Calcium and Phosphorous with concentration (ppm) of 344.37±6.24, 2.80±2.50, 35.60±6.50, 59.44±3.70 and 164.18±6.60, respectively with a trace amount of iron.

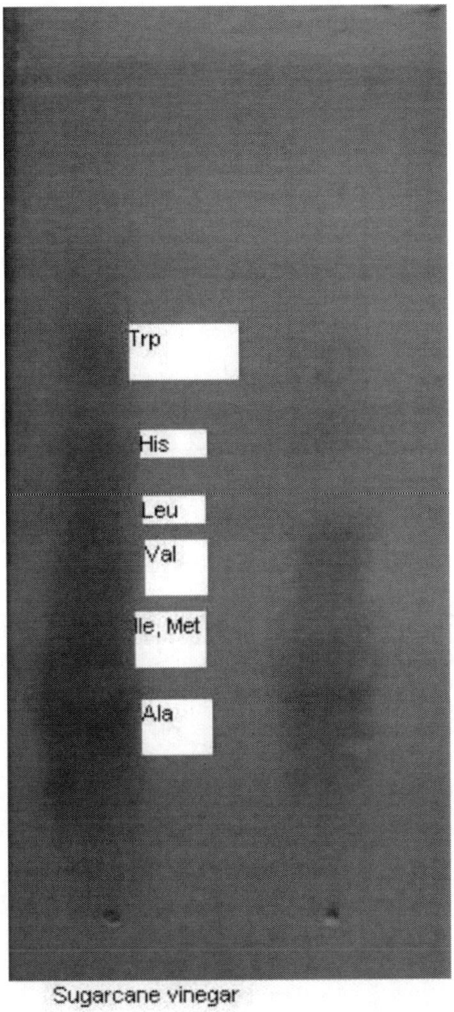

Figure 3. Amino acids analysis of sugarcane vinegar by Thin layer chromatography.

CONCLUSION

India consumes synthetic vinegar that has no health benefits, hence there is a need to promote health enriched fermented vinegar using widely available fermentable sugar rich sources such as sugarcane. The present study compared the batch and semicontinuous fermentation process of sugarcane vinegar production. A semi continuous type PVC acetifier containing wood shavings immobilized cells of *A. aceti* was designed and employed for the production of vinegar from sugarcane juice that produced 6.8% (w/v) volatile acidity in 3 days at 5 L scale. Vinegar produced by this technique yielded better body, color and odor thus proving that wood shaving's adsorbed bacterial biocatalyst contributes a good and effective support for bacterial vinegar fermentation. The present batch vinegar production technique is culture defined, microbiologically safe and cost effective fermentation and is also presented in Figure 4 as a flow chart.

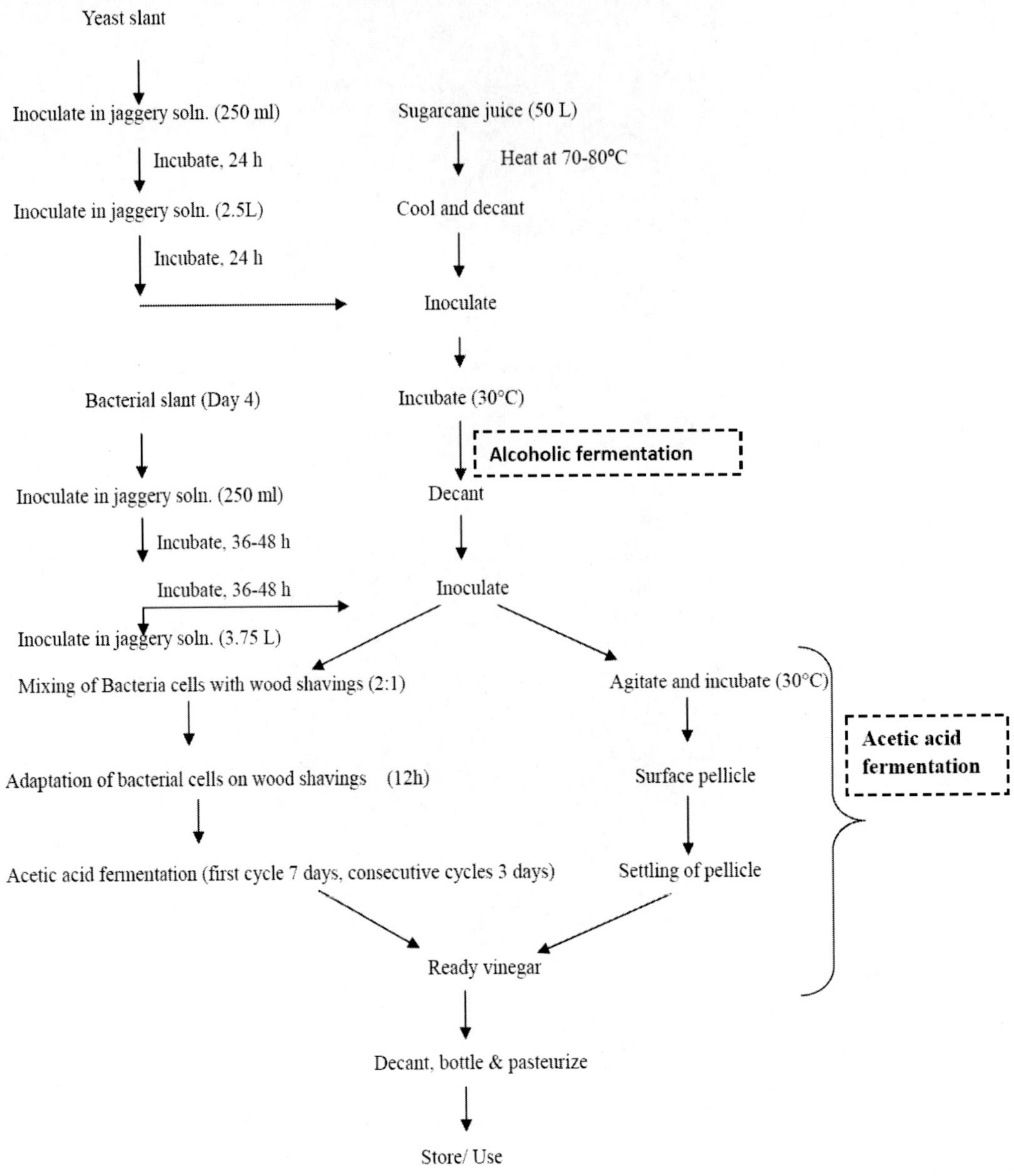

Figure 4. Flow sheet of sugarcane vinegar production.

ACKNOWLEDGMENTS

The authors thank Department of Science and Technology, New Delhi for financial assistance. The English editing carried out by Dr Ashu Toor, Assistant professor of English, Department of agricultural journalism, languages and culture is also duly acknowledged.

REFERENCES

[1] Kocher GS, Phutela RP, Dhillon HK, Uppal SK, Arora JK and Bakshi DK (2013). Standardization of an Economical Bioprocess for Production of Natural Vinegar from Sugarcane. *Sugar Tech,* DOI: 10.1007/s12355-013-0245-6.

[2] The Gazette of India. (1996). Ecomark criteria for food additives. Part II-Section 3(i), No. 170.

[3] Maal BK and Shafiee R (2010). Characterization of a *Acetobacter* strain isolated from Iranian peach that tolerate high temperature and ethanol concentrations. *International Journal of Medicine and Medicinal Science.* 1: 1-14.

[4] Kondo S, Tamaya K, Tsukamoto Y, Ikede K and Yamori Y (2001). Antihypersensitive effects of acetic acid and vinegar on spontaneously hypersensitive rats. *Bioscience, Biotechnology and Biochemistry.* 65: 2690-94.

[5] Sugiyama A, Saitoh M, Takahara A, Satoh Y and Hashimoto K (2003). Acute cardiovascular effects of a new beverage made of wine vinegar and grape juice, assessed using a *in vivo* rat. *Nutritional Research.* 23: 1291-96.

[6] Kuriowa T (1977). Rice vinegar - An oriental home remedy. 1st ed. *Pub Kenko Iga Kusha Co. Ltd.*

[7] Jimenez-Hornero JE, Santos – Duenas IM and Garcia-Garcia GI (2009). Optimization of biotechnological processes. The acetic acid fermentation. Part I: The proposed model. *Journal of Biochemical Engineering.* 45: 1-6.

[8] Lea AGH (1989). Cider vinegar. In: Downing, D.L. (Ed.) Processed apple products. New York: *Van Nostrand Reinhold.* 279–301.

[9] Fregapane G, Fernández HR, Salvador MD (2003). Continuous production of wine vinegar in bubble column reactors of up to 60-litre capacity. *European Food Resource Technology* 216:63–67.

[10] Sossou SK, Ameyapoh Y, Karou SD and Souza CD (2009). Study of pineapple peelings processing into vinegar by biotechnology. *Pakistan Journal of Biological Sciences* 11: 859-865.

[11] Brodelius P and Vndamme EJ (1987). Immobilized cell systems. In: Biotechnology., H.J. Rehm and G. Reed, Eds. *Verlag Chemie*: Germany 7(A): 405–464.

[12] Lotong N, Malapan W and Boongorsrang A (1989). Production of vinegar by acetobacter cells fixed on a rotating disc reactor. *Applied Microbiology and Biotechnology.* 32: 27-31.

[13] Miller CC, Park Y, Pariza MW and Cook ME (1994). Feeding conjugated linoleic acid to animals partially overcomes catabolic responses due to endotoxin injection. *Biochemical and Biophysical Research Communications* 198:1107-1112.

[14] Kocher GS, Kalra KL and Phutela RP (2006). Comparative production of sugarcane vinegar by different immobilization techniques. Institute of Brewing 112: 264-266.

[15] Park YS, Ohtake H, Fukaya M, Okumura H, Kawamura Y and Toda K (1989). Enhancement of acetic acid production in a yield cell density culture of *Acetobacter aceti. Journal of Fermentation Bioengineering* 68(5): 314-319.

[16] Nanba A, Kimura K and Nagai S (1985). Vinegar production by *Acetobacter rancens* cells fixed on a hollow fiber module. *Journal of Fermentation Technology* 63: 175–179.

[17] Ory I De, Romero LE and Cantero D (2004). Optimization of immobilization conditions for vinegar production. Siran, wood chips and polyurethane foam as carriers for *Acetobacter aceti. Process Biochemistry* 39(5): 547–555.

[18] Lashley D (1987). Alcoholic beverages making, composition and process. *US Patent* 4784859.

[19] Kida K, Asano S, Yamadaki M, Iwasaki K, Yamaguchi T and Sonoda Y (1990). Continuous high ethanol fermentation from cane molasses by flocculating yeast. *Journal of Fermentation and Bioengineering* 69(1): 39-45.

[20] Dombek KM, and Ingram LO (1987). Ethanol production during batch fermentation with *S. cerevisiae*: Change in glycolytic enzymes and internal pH. *Applied Environmental Microbiology* 53(6): 1286-1296.

[21] Fu N and Peiris P (2008). Cofermentation of a mixture of glucose and xylose to ethanol by *Zymomonas mobilis* and *Pachysolen tannophilus. World Journal of Microbiology and Biotechnology* 24: 1091-1097.

[22] Uppal S (2009). Co-fermentation of xylose and glucose by *Pachysolen tannophilus* and *Saccharomyces cerevisiae* to produce ethanol. M.Sc *Thesis, PAU, Ludhiana.*

[23] Sankhla S, Chaturvedi A, Kuna A and Dhanlakshmi K (2012). Preservation of Sugarcane Juice Using Hurdle Technology. *Sugarcane Technology* 14(1): 26-39

[24] Muhammad A, Ullah J, Muhammad A and Alam Z (2010). Evaluation of strawberry juice preserved with chemical preservatives at refrigeration temperature. *International Journal of Nutrition and Metabolism* 2(2): 27–32.

[25] Sacchi KL, Bisson LF and Adams DO (2005). A review of the effect of winemaking techniques on phenolic extraction in red wines. *American Journal of Ecology and Viticulture* 56:197–20.

[26] Plata C, Millan C, Mauricio JC and Ortega JM (2003). Formation of ethyl and isoamyl acetate by various species of wine yeasts. *Food Microbiology 20*: 217–224.

[27] Kocher GS and Dhillon HK (2013). Fermentative production of sugarcane vinegar by immobilized cells of *Acetobacter aceti* under packed bed conditions. *Sugar Tech* 15(1): 71-76.

[28] Caputi Jr. A and Wright D (1969). Collaborative study of the determination of ethanol in wine by chemical oxidation. *Journal of Association of Official Analytical Chemists* 52: 85–88.

[29] Amerine MA, Kunkee RE, Ough CS, Singleton VL and Webb AD (1980). The Technology of Wine Making. 4th ed. The AVI Publishing Co: *Westport, CT.*

[30] Malik CP and Singh M (1991). Plant enzymology and Histoenzymology. *Kalyani Publishers,* Delhi.

[31] AOAC (2000). Official methods of analysis, association of official analytical chemists, Washington, DC 18 ed. 940.13.

[32] Berraud C (2000). Production of highly concentrated vinegar in fedbatch cultures. *Biotechnology Letters* 22: 451–454.

[33] Kaur M and Kocher GS (2002). Ethanol production from molasses and sugarcane juice by an adapted strain of *Saccharomyces cerevisiae. Indian Journal of Microbiology*, 42 (3): 255-257.

[34] Castro-Martinez, Escudero-Abarca BI, Rodriguez JZ, Hyward-Jones PM and Agularuscanga MG (2005). Effect of physical factors on acetic acid production in Brettanonyces strains. *Journal of Food Processing and Engineering,* 28: 133-143.

[35] Gonzalez-Saiz MJ, Garrio-Vidal D and Pizarro C (2009). Modelling the industrial production of vinegar in aerated stirred fermentation in terms of process variables. *Journal of Food Engineering* 91:183-96.

[36] Kocher GS, Dhillon HK and Joshi N (2012). Scale up of sugarcane vinegar production by recycling of successive fermentation batches and its organoleptic evaluation. *Journal of Food Processing and Preservation.* DOI: 10.1111/jfpp.12050.

[37] Ferreira MJ, Swarnakar R and Saliva FLH (2003). Effect of nutrient sources on bench scale vinegar production using response surface methodology. R*evista Brasileira de Engenharia Agricola e Ambiental.* 9, 73-77.

[38] Kim J, Choo J, Wee Y, Yun J and Ryu H (2005). Culture medium optimization for acetic acid production by a Persimmon vinegar-derived bacterium. *Applied Biochemistry and Biotechnology,* 5:121-124.

[39] Gullo M and Giudici P (2008). Acetic acid in traditional balsamic vinegar, phenotypic traits relevant for starter cultures selection. International *Journal of Food Microbiology* 125: 46-53.

[40] Ebner H (1969). Process for acetic acid fermentation. *US Patent 3445245*

[41] Mehaia MA and Cheryan M (1991). Fermentation of date extracts to ethanol and vinegar in batch and continuous membrane reactors. *Enzymes and Microbial Technology* 13(3): 257–261.

[42] Garg N, Tandon DK and Kalra SK (1995). Production of mango vinegar by immobilized cells of *Acetobacter aceti. Journal of Food Science and Technology. Mysore,* 32(3): 216-218.

[43] Marques F P P (2010). Quality pattern and identity of commercial fruit and vegetable vinegar (Acetic acid fermentation). *Ciênc. Tecnol. Aliment.* 30(1): 119-126.

[44] Karmarkar R, Ghosh AK and Gangopadhya H (2010). Effect of Pretreatments on Physico-Chemical Characteristics of Sugarcane Juice. *Sugarcane Technology,* 13(1):47–50.

[45] Phanikumar HK (2011). Sugarcane juice powder by spray drying. *Science tech entrepreneur.* www.panelamonitor.org/documents/344/sugarcane-juice-powder-spray-drying-technique. Accessed 28 February 2012.

[46] Joshi N (2010). Studies on standardization of acetic acid fermentation conditions for production of concentrated vinegar from sugarcane juice. M.Sc. *Thesis, P.A.U,* Ludhiana.

[47] Tesfaye W, Morales ML, García-Prailla MC and Troncoso AM (2002). Wine vinegar: technology, authenticity and quality evaluation. *Trends in Food Science and Technology* 13:12–21.

In: Current Status of Sugarcane Research in India
Editors: A. K. Tiwari, M. Lal and A. K. Singh

ISBN: 978-1-63463-458-8

Chapter 7

DISEASE SCENARIO OF SUGARCANE SEEDLINGS AND STANDING CROPS IN BIHAR

Md. Minnatullah* and S. Dohare
Division of Plant Pathology, Sugarcane Research Institute, Pusa, Bihar

ABSTRACT

The frequency of fungal infection of sugarcane seedlings varied from 48.65 to 95.00 per cent and the mortality among the diseased seedlings ranged from 29.35 to 69.15 percent. *Aspergillus niger, Cladosporium cladosporioides, Drechslera rostrata* and *D. sacchari* were found associated only with seeds, whereas, *Diplodia macrospora, Alternaria alternata, Curvularia lunata, Fusarium semitectum* and *Drechslera halodes* were isolated both from affected seedlings and seeds. The fungi and their filtrates caused seedling mortality ranging from 7.27-34.78 and 3.33-40.00 percent, respectively. Fungal infestations also retarded the seed germinability. Adequate control of diseases caused by *H. halodes* and *A. alternata* was obtained when fungicides were applied at the time of sowing/emergence and after five days of emergence. Pre sowing soil treatment and seed treatment with Bavistin, Thiram, Captan or Brassicol were equally effective against root rot. Consequently the seed germination was also enhanced.

Red rot, smut and wilt diseases retarded the sett viability, cane length, weight, girth, juice and moisture contents, chlorophyll 'a' and 'b', number of internode, length & width of leaves. A differential decline in pol, brix, purity, commercial cane sugar and Jaggery recovery was also observed due to infection with these diseases. In infected cane juice the contents of boron, calcium, iron, sodium and nitrogen were increased while those of zinc, copper, magnesium, phosphorous, manganese and potassium were reduced. A considerable increase in the concentrations of amino acids and phenolic compounds was observed due to infection of red rot pathogen. Planting of sugarcane setts having different levels of red rot infection caused drastic reduction in sett germinability. The crop raised from naturally infected setts showed the lowest germination (19.5 %) and the highest red rot incidence (34.5 %) followed by the crop raised from artificially inoculated setts that showed (22.9%) germination and (15.5%) disease incidence. Maximum settling mortality (84.7%) was recorded, when both nodal and internodal infection occurred

* Corresponding author: Email: minat.pusa@gmail.com.

simultaneously, followed by (39.4%) when setts were taken from six month old inoculated cane, 35.3% with setts having nodal infection only and 32.5% in setts taken just before planting. Red rot fungus in association with wilt fungi led to considerable damage compared to red rot fungus alone.

The extent of losses in yield attributes and juice quality increased with the increasing level of red rot infection. Maximum losses were recorded at 100 % infestation level. Sett treatment with Bavistin (0.1 %) was found to be more effective than with *T. viride* and *T. harzianum* in enhancing the sett germinability and reducing red rot development. Similarly, sett treatment with 0.25 % Baylaton was found most effective in eradicating the external sett borne infection with smut disease. Groundnut & mustard cakes at 2% levels have effectively reduced the wilt pathogen population, disease incidence and improved the plant growth.

Keywords: Sugarcane seedlings, seedling blight, red rot, root rot, smut, wilt, diseases and management

INTRODUCTION

Diseases are one of the major constraints in the profitable cultivation of sugarcane. More than 25 diseases of sugarcane caused by fungi, bacteria, virus, mycoplasma and nematodes have been reported from Bihar. These diseases attack both seedlings as well as matured canes. In Bihar, root rot and blight are important fungal diseases of seedlings, whereas, red rot, smut and wilt are major fungal diseases of standing canes.

Propagation of sugarcane seedlings from true seeds (fuzz) is an essential step in the development of new commercial clones. Approximately 40,000 to 50,000 seedlings are raised every year at Sugarcane Research Institute (SRI), Pusa, Bihar. A severe seedling mortality of sugarcane was observed at SRI during February-March, 1986 in seedlings raised from the seed procured from SBI, Coimbatore.

The incidence of infection varied from 48.65 to 95.00 per cent and the mortality among the diseased seedlings ranged from 29.35 to 69.15 percent causing poor stand of seedlings. Fungal infection of the inflorescence leads to the production of diseased seeds, thus constituting a serious menace in hybridization and nursery raising [1].

Red rot, smut and wilt are important and destructive diseases of standing cane causing considerable damage to both yield and quality parameters. Bihar is considered to be hot spot for sugarcane diseases due to prevalence of congenial environment particularly temperature and humidity. Mortality of seedlings due to root rot disease and damage to standing canes due to red rot, wilt and smut diseases are of great concern in Bihar. It was therefore desirable to work out these diseases and their management.

DISEASES OF SEEDS, SEEDLINGS AND THEIR MANAGEMENT

A perusal of literature on seed pathology indicates that no sufficient attempts have been made to study the seedling mortality of sugarcane and its management in Bihar. Hence, to start with, diseased samples showing the characteristic symptoms of seedling mortality disease were collected from seed bed nursery of SRI, Pusa. Repeated isolations from the affected seedlings yielded *H. halodes* and *A. alternata*. On inoculation, these fungi produced two distinct types of disease symptoms which were usually observed simultaneously on the infected plants in nature as well.

Seedling Mortality

Fungal infection of sugarcane seedlings raised from true seed (fluff) has assumed considerable importance due to poor stand of seedlings in seed bed and thus constituting a serious menace in nursery raising. Kumar et al. [1] reported that the incidence of seedling infection varied from 48.65 to 95.00 per cent and mortality amongst the infected seedlings ranged from 29.35 to 69.15 per cent causing poor stand of seedlings (Table 1).

Table 1. Incidence of Seedling Mortality of Sugarcane

Crosses	Incidence %	Mortality %	Cross	Incidence %	Mortality %
Co 7314 GC	83.45	50.36	Co 1148 x Bo 34	60.00	37.86
CoL 9 GC	80.62	53.43	Co 671 x Co 419	81.25	59.12
Co 1148 GC	67.70	48.73	Co 740 x Co 671	83.75	61.73
CoJ 64 GC	92.50	63.73	Co 740 x Co 775	86.25	46.17
BO 17 GC	58.33	36.43	Co 449 x Co 1148	95.00	69.15
BO 99 GC	48.65	29.35	BO 47 x Co 775	75.00	43.36
Co 630 GC	82.50	49.39	BO 91 x Co 62399	65.00	31.95
Co 671 GC	83.53	53.78	Co 8213 x CoS 767	89.37	58.12
Co 740 x Co 6806	72.50	38.77	CoC 671 x Co 1148	84.53	51.46
Co 7201 x Co 775	77.50	50.12	BO 91 x Co 775	67.91	43.36
BO 17 x Co 1148	78.75	56.36			

Average of four replications.

Fungi Isolated from the Seed

Kumar et al. [1] reported that all 21 seed samples representing 13 crosses and 8 general collections yielded fungal growth. Fungal infestation in untreated seeds was 16.66%, whereas, it was only 11.05% in pretreated seeds (Table 2).

Table 2. Fungal Infection of Sugarcane Seeds

Crosses	Infestation %		Crosses	Infestation %	
	Untreated	Treated		Untreated	Treated
Co 7314 GC	21	13	Co 7201 x Co 775	17	11
CoL 9 GC	16	8	BO 17 x Co 1148	12	8
Co 1148 GC	18	11	Co 1148 x Bo 34	16	13
CoJ 64 GC	26	20	Co 671 x Co 419	14	9
BO 17 GC	11	6	Co 419 x Co 1148	25	16
BO 99 GC	9	7	Bo 47 x Co 775	11	7
Co 6304 GC	19	13	BO 91 x Co 62399	9	6
Co 671 GC	23	16	Co 8213 x Cos 767	15	9
Co 740 x Co 6806	13	9	Co 671 x Co 1148	18	13
Co 740 x Co 671	20	12	BO 91 x Co 775	13	6
Co 740 x Co 775	24	19	Mean	16.67	11.05

Average of four replications.

ECONOMIC REPERCUSSION

All the seed samples were found infected with *Aspergillus niger, Curvularia lunata* and *Alternania alternata, Fusarium semitectum and Helminthosporium halodes* and *Cladosporium cladosporioides* were isolated from three seed samples, whereas, *Drechslera sacchari* and *D. rostrata* from four samples [1]).

Among the fungi isolated from sugarcane true seeds, *Helminthosporium halodes* is a new report. These fungi produced adverse effect on seed viability. Minimum seed germination (23%) was recorded in seed inoculated with *Drechslera halodes* followed by 27% with *D. rostrata*, 29% with *D. Sacchari*, 30% with *Diplodia macrospora,* 35% with *Cladosporium cladosporioides,* 39% with *Fusarium semitectum,* 48% with *Alternania alternata,* 49% with *Curvularia lunata* and 55% with *Aspergillus niger* as against 63% in control. Maximum seedling mortality (34.78%) was noted due to *D. halode*s whereas, 33.33, 30.61, 29.62, 24.14, 20.61, 17.97, 11.42 and 7.27 percent seedling mortality was observed due to *D. macrospora, C. lunata, D. rostrata, D. Sacchari, F. semitectum, A. alternata, C. cladosporiodes,* and *A. niger,* respectively. Close examination of seedlings manifested blacking and rotting of collar region. Therefore, seedlings got wilted and finally dried (Table 3).

Table 3. Prevalence of seed mycoflora of sugarcane causing seedling mortality

Fungi isolated	Germination %	Mortality %
Alternaria alternate	48.00	17.97
Aspergillus niger	55.00	7.27
Curvulania lunata	49.00	30.61
Fusarium semitectum	39.00	20.61
Drechslera sacchari	29.00	24.14
Diplodia macrospore	30.00	33.33
Cladosporium cladosporioides	35.00	11.42
Drechslera rostrata	27.00	29.62
Drechslera halodes	23.00	34.78
Control	63.00	0.00

Average of four replications.

Table 4. Effect of culture filtrates of seed mycoflora of sugarcane on seedling mortality

Seed Mycoflora	Seedling Mortality % Concentration of filtrates				Mean
	0 %	25 %	50 %	100 %	
Alternaria alternata	0.00	0.00	10.00	20.00	10.00
Aspergillus niger	0.00	0.00	0.00	10.00	3.33
Curvulania lunata	0.00	10.00	30.00	40.00	26.66
Cladosporium cladosporioides	0.00	0.00	10.00	20.00	10.00
Diplodia macrospora	0.00	20.00	50.00	50.00	40.00
Drechslera rostrata	0.00	10.00	20.00	30.00	20.00
Drechslera halodes	0.00	20.00	30.00	40.00	30.00
Drechslera sacchari	0.00	10.00	20.00	30.00	20.00
Fusarium semitectum	0.00	20.00	30.00	30.00	26.66
Mean	0.00	10.00	22.22	30.00	-

Average of four replications.

Culture filtrates of isolated fungi also caused seedling mortality but to the varying extent [1]. Maximum seedling mortality (40%) was recorded with cultural filtrate of *D. macrospora* followed by filtrates of *D. halodes* (30%), *C. lunata* and *F. semitectum* (26.66%), *D. rostrata* and *D. sacchari* (20%), *A. alternata* and *C. cladosporioides* (10%) and *A. niger* (3.33%). The level of seedling mortality was increased with the increasing concentration of filtrates. Maximum mortality (30%) was recorded with 100% concentration, whereas, only 22.22 and 10.00 percent mortality was observed at 50 and 25 per cent concentrations of filtrates, respectively. Fungi of seed origin are, thus mainly responsible for poor germination of sugarcane seed (Fluff) and poor stand of seedlings (Table 4).

MANAGEMENT OF DISEASES OF SEEDLING

Seedling blight of sugarcane caused by *Helminthosporium halodes* and *Alternania alternata* is a serious disease causing considerable damage to the seedlings. Due to this disease, seedling population in seed bed nurseries is considerably declined causing problems to sugarcane breeders. Adequate control of this disease was obtained by Rai et al. [2] by spraying Bavistin (0.1%), Blitox-50 (0.25%), Ridomil MZ (0.1%), Indofil-45 (0.25%), Bordeaux mixture (1%) and Brassicol (0.25%) just after sowing of seeds, at emergence of seedlings and after 5 days of emergence. Spraying of these fungicides 10 days after emergence of seedlings, could not control the disease up to the mark. However, Ridomil MZ and Bavistin were found superior to other fungicides tested. Maximum control of the disease was recorded when fungicides were sprayed just after sowing of sugarcane seeds (Fluff) (Table 5 & 6).

Data presented in Table 7 clearly indicated that root rot disease of sugarcane seedling caused by *Pythium graminicolum* gave poor stand of seedlings in seed bed nursery. The extent of root rot disease was inversely correlated with age of seedlings.

Table 5. Effect of time of spraying with fungicides of seedlings blight caused by *Helminthosporium halodes*

Fungicides	Concentration (%)	Blight (%) Time of spraying at					
		0 day	5 days	10 days	15 days	20 days	Mean
Bavistin	0.10	6.0	9.5	13.5	45.0	60.5	26.9
Blitox – 50	0.25	16.0	18.5	20.0	55.0	75.0	36.9
Indofil M-45	0.20	15.0	15.0	20.5	50.5	70.0	34.2
Bordeaux mixture	0.25	20.0	22.5	25.5	60.0	85.5	42.7
Ridomil MZ	0.10	8.5	10.0	15.0	35.5	65.0	26.8
Brassicol	0.25	18.5	20.0	21.5	50.5	75.0	37.1
Mean	-	14.0	15.9	19.3	49.4	71.8	-
Control	-	82.4	88.5	90.7	94.2	100.00	-

Average of two replications.

C. D. at 5% for time spary = 1.95, treatment = 2.30 and interaction = 5..15.

Table 6. Evaluation of fungicides against seedling blight caused by *Alternaria alternate*

Fungicides	Concentration (%)	Blight (%) Time of spraying					
		0 day	5 days	10 days	15 days	20 days	Mean
Bavistin	0.10	5.0	8.5	10.5	45.0	50.0	23.7
Blitox – 50	0.25	20.0	22.5	25.0	50.5	70.0	37.6
Indofil M-45	0.20	15.0	18.5	20.5	55.0	75.0	36.8
Bordeaux mixture	0.25	22.5	25.0	25.0	60.5	80.0	42.6
Ridomil MZ	0.10	5.5	6.0	10.5	40.5	50.5	22.6
Brassicol	0.25	20.0	22.5	22.5	50.0	75.0	38.0
Mean	-	14.6	17.16	18.91	50.25	66.75	-
Control	-	91.25	93.80	95.1	99.2	100.0	-

Average of two replications.
C. D. at 5% for time spary = 1.93, treatment = 2.88 and interaction = 5.11.

As the age of sugarcane seedlings increased, the extent of root rot disease decreased. Seedlings over 3 months old could not be infected. Pre-sowing soil treatments or seed treatments with Bavistin (0.1%), Thiram, Captan and Brassicol (0.25%) were found equally effective in enhancing seed germination. However, Topsin-M and Ronilan (0.1%), Emisan-6 and Ceresan wet (0.25%), Bordeaux mixture and chestnut compound (1%) had adverse effect on seed germination. The reduction in germination due to these fungicides was higher in pre-sowing seed treatment as compared with pre-sowing soil treatment. Although, these fungicides also proved efficacious in reducing the root rot incidence but differed in their efficacy. Soil drenching prior to sowing as well as post emergence were superior to pre-sowing seed treatment in checking the extent of disease [3].

1. Pre sowing soil treatment.
2. Pre sowing seed treatment.
3. Post emergence soil treatment.

Table 7. Effect of fungicides on seed germination and root rot disease of sugarcane seedlings

Fungicide	Germination %		Root rot %			Control %		
	1	2	1	2	3	1	2	3
Thiram	59.5	62.6	11.6	18.6	13.4	77.5	64.4	75.6
Captan	87.6	59.9	16.7	21.7	17.5	67.3	58.4	68.1
Brassicol	56.8	57.2	9.3	21.8	11.8	81.8	58.2	78.5
Bavistin	61.3	61.8	46.8	50.4	47.3	8.4	3.5	13.7
Topsin-M	46.5	48.5	21.4	28.6	24.0	58.1	45.2	56.2
Ronilan	41.3	44.6	26.3	33.3	29.6	48.5	36.2	46.0
Kitazin	35.9	38.6	18.2	26.3	20.6	64.4	49.6	62.4
Emisan-6	28.9	19.8	7.8	38.6	11.8	84.7	26.1	78.5
Cersan wet	26.7	16.4	9.7	36.9	11.7	81.0	29.3	78.6
Bordeaux mixture	31.4	21.8	5.8	29.3	8.3	88.7	43.9	84.9
Cheshnut compound	27.5	27.8	6.2	31.7	8.9	87.9	39.3	83.8
Control	35.6	38.1	51.1	52.2	54.8	-	-	-
C. D at 5%	5.2	5.9	7.8	6.7	7.1	-	-	-

RED ROT

Red rot is the most destructive disease in Bihar. Since several high yielding and high sugared varieties were wiped out from the cultivation due to epidemic of this disease, a detailed study was made on economic repercussions, variability and management of this disease.

VARIABILITY IN RED ROT PATHOGEN

In nature, the development of new pathogenic strains/races of *C. falcatum* has been reported from time to time due to which resistant varieties collapsed disturbing the entire varietal setup in Bihar.

Knowledge on variation in *C. falcatum* Went and the incidence of red rot disease of sugarcane is the pre-requisite for evolving red rot resistant sugarcane varieties. Physiological, cultural and pathological characteristics can be used to differentiate distinct races. Kumar et al. [4] categorized 80 isolates of red rot pathogen into 7groups on the basis of growth rate, colony characters, extent of sporulation and spore size (Table 8).

Table 8. Morphological and Cultural Variation among isolates of *Colletotrichum falcatum*

Isolates	Colony diameter in mm	Colony characters	Sporulation per ml.	Size of Conidia
1.	80	Colony uniformly circular, compact, margin zig-zag, cottony white in colour which turned grey with age. Acervuli brown in colour.	76×10^2	18-36 μ x 5.7 μ (24.5 μ x 5.6 μ)
2.	76	Colony circular, semi compact, margin smooth and entire, whitish in colour. Acervuli deep brown in colour.	79×10^2	21-37 μ x 5.7 μ (28.7 μ x 5.2 μ)
3.	85	Colony uniformly circular, compact in texture, margin smooth, grey in colour, Acervuli black in colour.	116×10^2	20-26 μ x 4-7 μ (23.7 μ x 4.6 μ)
4.	90	Colony circular, margin zig-zag, loose with sparse aerial mycelium, whitish in colour. Acervuli light brown in colour.	110×10^2	23-25 μ x 5.6 μ (26.3 μ x 5.2 μ)
5.	71	Colony circular, margin smooth and undulating, whithish grey in colour, loose fluffy. Acervuli deep brown in colour.	96×10^2	28-40 μ x 5.7 μ (34.4 μ x 5.8 μ)
6.	74	Cottony growth of mycelium which turns grey with age, margin zig-zag, loose in texture. Acervuli deep brown in colour.	86×10^2	25-40 μ x 5.6 μ (29.7 μ x 5.6 μ)
7.	98	Colony circular margin smooth and entire, dirty white in colour, loose. Acervuli black in colour.	126×10^2	16-26 μ x 5.6 μ (21.5 μ x 5.3 μ)

On the basis of reaction of 1-7 isolates on 4 sugarcane varieties namely BO 11, Co 453, BO 34 and BO 43, isolate 07 appeared to be the most virulent because it produced susceptible reaction even in resistant varieties (BO 34 and BO 43), whereas, the remaining isolates produced highly resistant to moderately resistant reaction on the same set of varieties.

Table 9. Pathological Variation among Isolates of *Colletotrichum falcatum*

isolates	Linear spread of red rot lesions				Mean
	BO 11	Co 453	BO 34	BO 43	
1	34.1	32.6	15.9	14.4	24.25
2	37.0	38.2	19.6	18.7	28.38
3	41.4	48.6	25.3	26.3	35.40
4	43.4	46.1	25.8	22.3	34.40
5	8.3	7.5	3.9	4.8	6.12
6	9.4	10.0	3.7	5.6	7.18
7.	57.1	69.8	49.3	48.6	56.2

The isolate (07) was prevalent in Riga, Lohat and Pusa cane growing belts. Isolates 5 and 6 showed a highly resistant reaction even in susceptible varieties (BO 11 and Co 453), whereas, isolate 07 had produced a highly susceptible reaction on these varieties. This indicated that isolates 5 and 6 differed from isolate 07 in pathological behaviour. The isolates (5 and 6) are distributed in Bihta and Warsaliganj areas of South Bihar. Isolates 1, 2, 3 and 4 showed resistant to moderately resistant reaction on resistant varieties (BO 43 and BO 34) and moderately susceptible to susceptible reaction on susceptible varieties (BO 11 and Co 453).

Hence, the isolates (1-4) also differed from isolates 5 and 6 as well as 7. These isolates (1-4) are distributed in major cane growing areas of North Bihar. On the basis of above observations, it can be concluded that at least 3 strains of red rot pathogen are prevalent in Bihar (Table 9).

The most virulent strain (isolate 07) is distributed in Pusa, Lohat and Riga regions, whereas, the weakest strain (isolates 5 and 6) is prevalent in sugarcane growing belt of south Bihar. The third strain having intermediate virulence (isolates 1, 2, 3 and 4) is prevalent in major cane growing areas of North Bihar. Kumar et al. [5] and Minnatullah & Kumar [6] also confirmed the presence of three pathotypes of red rot fungus in Bihar.

ECONOMIC REPERCUSSION

Isolates of red-rot pathogen (*Colletotrichum falcatum* Went.) caused considerable losses both in cane yield and sugar recovery through poor germination and impaired juice quality. Kumar et al. [7] reported that the reaction of isolates (I_1-I_{10}) on varieties BO 70 and BO 74 showed a differential interaction between varieties and isolates of red rot pathogen. On the basis of disease reaction on variety BO 70 isolates I_2 and I_6 appeared to be the most virulent followed by isolates I_8, I_7, I_1 I_9, I_3, I_4, I_5, I_3 and I_{10}. In case of variety BO 74, isolate I_9 was the most virulent followed by isolates I_4, I_8, I_2, I_7, I_{10}, I_1, I_3 and I_6. A considerable reduction in cane length, girth, weight and juice content of clones BO 70, and BO 74 due to isolates of red rot pathogen was recorded.

However, the extent of reduction appeared to be proportional to the virulence of isolates. Infection by red rot pathogen thus retarded the growth and development attributes of cane finally resulting into reduction in yield vis-à-vis juice content (Table 10).

A differential decline in pol, brix, purity, commercial cane sugar and jaggery recovery was also observed in response to red rot infection which appeared to be proportional to the virulence of these isolates. The fall in pol and brix was much higher as compared to purity. This suggested that impaired sucrose synthesis associated with increase in moisture content might be the major factor for deterioration in juice quality due to inversion of sucrose (Table 11).

In red rot infected cane juice, the contents of boron, calcium, iron, sodium and nitrogen were accumulated, whereas, zinc, copper, magnesium, phosphorus, manganese and potassium were depleted. The accumulation/depletion of these nutrients was directly proportional to the virulence of isolates of red rot pathogen and caused a disturbance in the nutrient balance in the cane juice (Table 12 & 13).

A considerable increase in the concentration of amino acid in cane leaves was observed by Kumar et al. [8] due to infection with isolates of red rot pathogen (*Collectotrichum falcatum* Went) (Table 14).

Kumar et al. [8] further reported that isolates of red rot pathogen also induced accumulation of total phenols both in cane stalks and juice (Table 15).

Red rot (*Collectotrichum falcatum* Went), being one of the most serious diseases of sugarcane, is known to occur in Bihar causing considerable losses both in yield and juice quality. Pandey et al. [9] observed that the planting of red rot infected sugarcane setts caused drastic reduction in sett germinability.

The extent of reduction varied according to the type of infection. The crop raised from naturally infected setts had the least germination (19.5%) and the highest red rot incidence (34.5%), followed by the crop of artificially inoculated setts having 22.9% germination and 15.5% disease incidence (Table 16).

Maximum settling mortality (84.7%) was recorded when both nodal and internodal infection occurred simultaneously, followed by 39.4% when setts were taken from 6 months old inoculated cane, 35.3% with setts having nodal infection only and 32.5% when setts were inoculated just before planting (Table 17).

Red rot fungus (*C. falcatum*) in association with wilt fungi (*Fusarium moniliforme* var. subglutinans and *Cephalosporium sacchari*) led to considerable damage as compared to red rot fungus alone (Table 18).

The reduction in cane weight, juice content, brix, sucrose content, purity and recovery of jaggery varied according to the levels of red rot infection. Data presented in Table 19 revealed that the extent of losses in yield attributes and juice quality increased with the increasing level of red rot infestation. Maximum losses in weight of cane, juice content, brix, sucrose content, purity and jaggery recovery were recorded at 100% infestation level.

Kumar [10], Kumar et al. [8, 11, 12, 13, 14], Minnatullah et al. [5, 16, 17] reported adverse effects on yield parameters, juice quality and nutrient contents in juice due to red rot infection.

Table 10. Effect of red rot isolates on cane yield attributes

Attributes	Varieties	Isolate										control	CV (%)	CD at 5%.
		I_1	I_2	I_3	I_4	I_5	I_6	I_7	I_8	I_9	I_{10}			
Disease reaction (0-9 scale)	BO 70	7.0	7.8	6.7	6.3	5.2	7.8	7.2	7.6	7.0	5.1	0.0	8.05	1.56
	BO 74	6.6	7.1	6.4	7.6	7.3	6.2	7.0	7.3	8.1	7.0	0.0	6.85	1.36
Cane length (cm)	BO 70	213.6	206.8	216.9	219.6	338.6	208.8	211.6	204.6	216.8	228.0	244.5	18.06	NS
	BO 74	212.3	216.7	230.4	201.8	203.6	224.3	209.6	207.5	200.3	218.6	235.7	16.34	NS
Cane girth (cm)	BO 70	2.10	230.4	2.11	2.19	2.31	1.87	1.96	1.67	1.81	2.23	2.45	14.06	NS
	BO 74	2.14	2.11	2.18	1.85	1.71	2.10	1.76	1.81	1.61	1.64	2.34	13.16	NS
Weight of millable canes (kg)	BO 70	8.9	2.18	8.6	9.4	9.8	7.8	8.8	8.0	8.6	9.6	12.6	10.26	2.92
	BO 74	8.5	8.6	8.3	6.3	6.4	7.6	6.1	6.8	5.9	6.6	9.8	11.08	2.52
Juice content (%)	BO 70	54.8	8.3	55.0	56.3	58.5	52.7	54.6	53.4	55.1	58.6	65.5	4.46	7.87
	BO 74	52.5	55.0	53.7	52.3	52.8	55.1	54.6	52.6	49.6	53.6	63.4	5.37	9.14

Table 11. Effect of red rot isolates on cane juice quality

Juice quality	Variety	Isolates										Control	CV (%)	CD at 5%.
		I_1	I_2	I_3	I_4	I_5	I_6	I_7	I_8	I_9	I_{10}			
Pol (%)	BO 70	14.7	13.2	15.0	15.2	15.8	12.9	14.6	14.2	15.0	15.3	17.7	3.48	1.63
	BO 74	14.9	14.9	15.6	13.0	13.4	15.8	15.0	13.6	12.5	15.0	17.9	4.06	1.88
Brix (%)	BO 70	17.7	16.8	17.9	18.0	18.6	17.4	17.6	16.9	17.8	18.4	20.5	3.66	2.05
	BO 74	17.8	17.6	18.1	16.2	16.6	18.3	17.6	16.5	15.0	17.4	20.0	2.60	1.42
Purity (%)	BO 70	83.05	78.57	83.80	84.44	84.95	74.14	82.96	84.02	84.27	83.15	86.34	2.44	6.36
	BO 74	83.71	84.16	86.18	80.25	80.72	86.33	85.22	82.42	83.33	86.21	89.50	2.69	7.15
CCS (%)	BO 70	9.23	7.98	9.49	9.66	10.08	7.46	9.17	9.00.	9.52	9.63	11.41	4.39	1.29
	BO 74	9.42	9.49	10.05	7.98	8.26	10.19	4.59	8.51	7.88	9.66	11.66	5.01	1.47
Jaggery recovery (%)	BO 70	9.2	8.4	9.4	9.6	9.5	8.6	8.7	8.1	9.3	9.8	11.8	9.72	2.85
	BO 74	9.6	8.8	9.5	8.4	8.9	9.3	9.1	8.6	7.9	9.3	10.9	6.44	1.85

Table 12. Accumulation of mineral nutrients (meq/100 ml cane juice) in red rot infected cane juice

Attributes	Varieties	Isolate										Control	CV (%)	CD at 5%.
		I_1	I_2	I_3	I_4	I_5	I_6	I_7	I_8	I_9	I_{10}			
Boron	BO 70	5.33	6.25	5.56	5.42	4.97	6.35	5.75	6.35	5.48	4.80	4.50	3.60	0.63
	BO 74	4.42	4.65	4.31	5.00	5.10	4.97	4.53	4.97	5.42	4.69	3.65	5.27	0.77
Iron	BO 70	8.49	9.30	8.89	8.79	8.39	8.99	9.40	8.99	8.68	8.28	6.50	3.55	0.98
	BO 74	7.87	7.78	6.97	8.49	8.59	8.18	7.58	8.18	8.69	8.28	6.59	2.28	0.56
Calcium	BO 70	2.46	2.50	2.40	2.35	2.25	2.60	2.59	2.60	2.48	2.15	1.30	6.08	0.45
	BO 74	2.33	2.29	2.06	2.42	2.52	2.28	2.15	2.28	2.59	2.36	1.60	5.41	0.38
Sodium	BO 70	0.653	0.667	0.565	0.543	0.523	0.783	0.588	0.783	0.591	0.509	0.417	7.58	0.413
	BO 74	0.886	0.836	0.813	0.913	0.937	0.861	0.793	0.861	0.967	0.928	0.583	6.94	0.185
Nitrogen	BO 70	585.0	630.0	635.0	573.0	541.0	616.0	0.591	616.0	531.0	598.0	371.0	7.34	132.36
	BO 74	563.0	605.0	551.0	685.0	631.0	618.0	0.556	618.0	701.0	585.0	406.0	9.69	123.67

Table 13. Depletion of mineral nutrients (meq/100 ml cane juice) in red rot infected cane juice

Attributes	Varieties	Isolate										Control	CV (%)	CD at 5%.
		I_1	I_2	I_3	I_4	I_5	I_6	I_7	I_8	I_9	I_{10}			
Zinc	BO 70	1.37	1.08	1.72	1.77	1.88	1.16	1.42	1.26	1.64	1.93	2.23	5.60	0.28
	BO 74	1.88	1.55	1.93	1.30	1.49	2.05	1.68	1.37	1.00	1.84	2.50	6.67	0.36
Manganese	BO 70	5.21	3.34	5.73	5.95	6.18	4.17	4.27	4.38	5.24	6.15	7.08	5.56	0.92
	BO 74	5.11	4.59	5.21	3.75	4.35	5.38	4.90	4.17	3.44	5.00	6.98	8.90	1.35
Copper	BO 70	1.13	0.98	1.31	1.38	1.46	0.86	1.19	0.91	1.28	1.26	1.92	6.38	0.25
	BO 74	1.38	1.13	1.46	0.93	1.06	1.33	1.21	0.96	0.84	1.32	1.96	8.23	0.32
Magnesium	BO 70	3.00	2.25	3.35	3.50	3.63	2.35	2.50	2.75	3.25	3.75	4.00	6.30	0.62
	BO 74	4.26	3.90	4.35	3.55	3.75	4.39	4.10	3.63	3.45	4.21	5.56	8.41	1.09
Potassium	BO 70	6.49	5.92	6.91	6.87	6.94	6.21	6.33	6.35	6.77	6.91	7.76	4.32	0.91
	BO 74	6.93	6.61	7.10	6.31	6.56	7.05	6.73	6.39	6.21	6.81	7.86	3.56	0.76
Phosphorus	BO 70	245	187	275	287	293	200	214	206	263	307	351	6.05	49.00
	BO 74	308	268	317	225	255	337	236	283	265	396	405	8.13	73.00

Table 14. Pathogenic behaviour of red rot isolates and their effect on free amino acidcontent in cane leaves

Isolates	Disease index (0-9)					Free amino acid (%)				
	BO 70	BO 74	BO 91	BO 109	Mean	BO 70	BO 74	BO 91	BO 109	Mean
RR_1	8.7	8.3	2.9	3.8	5.9	0.296	0.256	0.238	0.260	0.263
RR_2	8.0	8.3	1.3	2.8	5.1	0.269	0.226	0.227	0.231	0.238
RR_3	6.9	6.3	2.6	1.9	4.4	0.239	0.210	0.198	0.193	0.210
RR_4	5.7	5.9	1.9	1.6	3.8	0.246	0.198	0.205	0.189	0.210
RR_5	8.6	8.9	3.4	3.9	6.2	0.298	0.253	0.245	0.251	0.262
RR_6	7.6	7.8	1.8	2.6	5.0	0.259	0.250	0.226	0.216	0.238
RR_7	7.6	7.3	2.6	2.5	5.0	0.248	0.230	0.215	0.206	0.225
RR_8	8.2	8.6	3.1	3.6	5.9	0.289	0.241	0.250	0.245	0.256
RR_9	6.1	5.8	1.6	3.1	4.2	0.236	0.189	0.210	0.185	0.205
RR_{10}	7.8	7.5	1.3	3.0	4.9	0.278	0.233	0.221	0.225	0.239
Mean	7.5	7.5	2.3	2.9	-	0.266	0.229	0.224	0.220	-
Control	0.0	0.0	0.0	0.0	0.0	0.169	0.140	0.137	0.146	0.148

Table 15. Effect of red rot isolates on phenolic content in cane and its juice

Isolates	Total phenol (µg/g fresh tissues/juice									
	BO 70		BO 74		BO 91		BO 109		Mean	
	Stalk	Juice	Stalk	Juice	Stalk	Juice	Stalk	Juice	Stalk	Juice
RR_1	82.5	130.5	60.8	100.0	69.4	102.9	74.4	120.6	71.8	113.5
RR_2	76.4	120.8	58.3	88.7	63.9	99.6	71.6	112.7	67.6	105.3
RR_3	73.9	113.9	55.8	85.7	62.2	92.6	66.7	107.9	64.7	100.3
RR_4	78.2	110.8	55.8	84.9	62.5	88.9	65.9	109.4	65.1	98.5
RR_5	78.4	128.3	60.2	99.3	68.7	109.8	75.9	118.5	70.8	114.0
RR_6	75.8	116.5	56.4	85.8	63.3	99.9	70.2	110.9	66.4	103.3
RR_7	74.3	118.9	57.9	86.9	65.8	98.2	68.8	108.7	66.7	103.2
RR_8	80.3	126.4	60.6	96.7	69.6	109.5	72.9	120.9	70.8	113.4
RR_9	72.8	115.6	55.9	86.2	61.9	89.9	64.9	103.6	63.8	98.8
RR_{10}	76.8	119.6	56.8	90.8	69.8	98.8	70.8	114.6	68.6	106.0
Mean	76.9	121.1	57.6	90.7	65.7	99.0	70.2	113.1	-	-
Control	70.6	107.6	52.4	80.6	58.6	87.6	63.7	96.8	61.4	93.6

Table 16. Effect of red rot infected cane setts on sett germinability and disease development

Treatment	Variety	Germination (%)	Mean	Red rot (%)	Mean
Naturally infected setts	BO 70	18.5		30.6	
	BO 74	20.4	19.5	38.4	34.5
Artificially inoculated setts	BO 70	21.6		12.4	
	BO 74	24.3	22.9	18.4	15.5
Healthy setts	BO 70	32.6		0.0	
	BO 74	38.7	35.6	0.0	0.0

Table 17. Effect of types of red rot infection on settling mortality

Type of infection	Variety	Mortality (%)	Mean
Setts inoculated just before planting.	BO 70	28.5	
	BO 74	36.5	32.5
Setts taken from six months old inoculated cane	BO 70	35.4	
	BO 74	43.3	39.4
Nodal infection	BO 70	32.1	
	BO 74	38.5	35.3
Setts having both nodal and internodal infection	BO 70	75.8	
	BO 74	87.6	84.7

Table 18. Effect of wilt fungi on red rot development

Fungi	Red rot intensity (0-9 scale)		
	BO 70	BO 74	Mean
Colletotrichum falcatum	6.1	7.8	6.9
C. falcatum + *Cephalosporium sacchari*	6.8	8.1	7.4
C. falcatum + *F. moniliforme var. subglutinans*	7.5	8.8	8.1

Table 19. Extent of losses at different levels of red rot infection

Treatment	Percentage of losses in					
	Cane weight	Juice content	Brix	Pol	Purity	Jaggery (Gur)
100 H* + 0 D*	0.0	0.0	0.0	0.0	0.0	0.0
90 H* + 10 D*	3.6	4.5	1.2	1.8	3.2	0.8
80 H* + 20 D*	5.5	10.5	1.8	3.6	7.7	1.5
70 H* + 30 D*	8.6	16.3	2.1	4.8	12.7	2.4
60 H* + 40 D*	13.8	23.4	2.6	5.8	17.6	3.9
50 H* + 50 D*	16.2	32.3	3.1	6.8	19.3	5.3
40 H* + 60 D*	19.3	36.9	4.3	8.5	24.2	5.7
30 H* + 70 D*	25.6	41.2	5.4	8.9	29.6	6.3
20 H* + 80 D*	34.8	48.3	8.7	9.6	34.2	6.9
10 H* +90 D*	40.3	53.6	11.3	10.5	40.1	8.5
0 H* + 100 D*	53.6	59.3	14.8	11.6	46.3	9.8

*H = Healthy.

* D = Diseased.

Table 20. Effect of micronutrients and their concentration on growth and sporulation of *C. falcatum*

Treatment	Concentration (%)	Growth (mg)	Sporulation (100/ml)
Boric acid	100	169.50	2.65
	150	125.78	1.98
	200	32.03	0.58
	250	0.00	0.00
Ammonium molybdate	100	183.23	2.43
	150	109.55	0.90
	200	57.18	0.56
	250	0.00	0.00
Zinc sulphate	100	156.83	2.73
	150	59.00	0.53
	200	31.50	0.33
	250	0.00	0.00
Copper sulphate	100	139.93	2.13
	150	52.93	0.63
	200	0.00	0.00
	250		
Ferrous sulphate	100	150.80	2.40
	150	70.10	1.15
	200	19.85	0.70
	250	0.00	0.00
Manganese sulphate	100	136.73	2.28
	150	42.53	0.43
	250	0.00	0.00
Control	-	245.95	6.30

Table 21. Effect of micronutrient spray on cane growth and red-rot development

Treatment	Cane height (cm)	Red-rot infection (0-9 scale)				
		Condition to tops	Lesion width	While spot	Nodal transgression	Total
Manganese sulphate	187.2	0.5	1.6	1.1	1.6	4.8
Zinc sulphate	183.3	.07	1.8	1.3	1.9	5.7
Agromin	194.4	0.5	1.5	1.0	1.7	4.7
Chilamin	186.5	0.6	1.7	1.4	1.8	5.5
Control	169.2	0.8	2.4	1.8	2.7	7.7
C.D at 5%	9.4	-	-	--		0.668

MANAGEMENT

Kumar et al. [18] observed that copper and manganese were equally effective in inhibiting the growth and sporulation of red rot pathogen *(Colletotrichum falcatum* Went). They found that copper at 200 ppm, whereas, zinc, iron, boron and molybdenum at 250 ppm totally inhibited the growth and sporulation of *Colletotrichum falcatum* Went *in vitro* (Table 20). When manganese sulphate, zinc shuphate, Chilamin (chelated zinc) and Agromin, a mixture of iron (5.2%), manganese (3.2%), zinc (5.3%), copper (2.5%), boron 0.007% and molybdenum (0.002%) were sprayed at 1% concentration four times at 15 days interval starting from 15th June, the length of canes was considerably increased. Spraying with Agromin proved superior to other treatments in enhancing the cane height (194.4 cms) as against 169.3 cm in control. A critical perusal of the effect of these treatments on the development of the red rot disease, indicated that all the treatments significantly reduced the development of red rot lesions but agromin and manganese sulphate proved equally superior to chilamin and zinc sulphate (Table 21).

Analysis of juice samples obtained from healthy and infested canes revealed that contents of sucrose, manganese, zinc and phosphorus decreased while iron content increased due to disease infestation resulting in poor juice quality. The contents of chlorophyll a and b were also declined due to infection. However, after spraying of nutrients on cane crop, the juice quality of both healthy and inoculated canes was considerably improved. Contents of chlorophyll a and b were also increased (Table 22).

Kumar et al. [5] observed that sett treatment with 0.1% fungicidal solution enhanced the sett germinability, suppressed the growth and sporulation of red rot pathogen and reduced the incidence of red rot disease (Table 23 & 24).

Maximum sett germination (69.6%) was recorded with Bavistin followed by Jkstein (68.0%), Derosal (67.6%), Topsin-M (66.3%), Ronilan (62.6%), Kitazin and Agrozim (60.3%) and Saprol (59.6%) as against 53.3% in control. No growth of red rot pathogen (*C. falcatum*) was recorded in medium containing Bavistin, Jkstein, Saprol, Derosal Agrozim and Topsin-M even at 100 ppm concentration. However, slight growth of fungus was observed in medium containing Kitazin and Ronilan at similar concentrations as against 81.8 mm in control. Sett treatment with these fungicides also reduced the incidence of red rot disease. Minimum disease incidence (21.3%) was recorded when setts were treated with Bavistin followed by Derosal (23.8%), Jkstein (26.8%), Agrozim (31.2%), Topsin-M (32.3%), Saprol (36.6%), Ronilan (37.4%) and Kitazin (38.7%) as against 58.6% in control. Minnatullah, et al. [19] also reported that Bavistin was found to be more efficacious than *T. viride*. and *T. harzianum* in enhancing the sett germinability and in reducing red rot development. Maximum increase in germination (42.94%) was observed when the setts were treated with bavistin. Maximum reduction in settling mortality (42.39%) was recorded with *T. harzianum* whereas, maximum reduction in development of red rot disease (45.38%) was observed when the setts were treated with bavistin (Table 25).

Smut

Smut of sugarcane (*Ustilago scitaminea*) is one of the destructive diseases causing considerable losses both in yield parameters and juice quality. A severe epidemic of smut disease occurred during 1942-43 in Bihar affecting 66% of cane areas. Since then, this disease is perpetuating year after year causing considerable losses both in yield and juice quality.

Economic Repercussions

Kumar et al. [3] reported that in smutted canes of BO 110, CoJ 64, CoS 767 and Co 1158 varieties, the length, girth, numbers of internodes, weight and moisture, juice content in canes, as well as length and width of the leaves were considerably reduced.

Table 22. Effect of micronutrient spray and red-rot infection on juice composition and chlorophyll contents

Juice constituent	Manganese sulphate		Zinc sulphate		Agromin chilamin		Chilamin		Control		CD at 5%
	H	D	H	D	H	D	H	D	H	D	H
Pol (%)	16.70	15.00	16.10	14.10	17.00	15.50	16.30	14.40	15.80	13.50	0.620
Brix (%)	19.30	18.10	18.80	17.20	19.60	18.60	19.00	17.50	18.60	16.70	0.742
Purity (%)	86.50	82.80	85.70	81.80	86.70	83.30	85.90	82.10	84.90	80.80	0.748
Invertase activity Neutral	5.60	7.37	5.62	7.59	5.65	7.18	5.64	7.48	5.88	8.16	0.536
(g/100 ml juice) Acidic	4.27	4.86	4.35	5.00	4.23	4.80	4.31	4.93	4.85	5.32	0.484
Chlorophyll (mg/g leaf)	a 0.53	0.43	0.49	0.38	0.54	0.45	0.50	0.39	0.48	0.33	0.045
Chlorophyll (mg/g leaf)	b 0.60	0.46	0.53	0.38	0.58	0.48	0.54	0.40	0.52	0.32	0.049
Manganese (p/m)	2.11	1.89	1.82	1.49	1.93	1.69	1.89	1.60	1.85	1.48	0.104
Zinc (p/m)	5.85	5.65	6.30	6.18	6.35	6.25	6.25	6.10	5.53	5.20	0.067
Iron (p/m)	24.63	28.53	23.87	28.75	26.75	30.10	24.13	28.81	22.10	28.63	1.299
Phosphorus (p/m)	372.50	335.00	360.00	308.00	377.00	352.50	362.00	371.50	350.00	296.50	3.851

Table 23. Effect of fungicides (L) as sett dresser (L) on sett germinability and red rot infection

Treatment	Sett germinability (%)						Red-rot incidence			
	Treatment period in minute									
	0	15	30	45	60	0	15	30	45	60
Bavistin	44.6	52.3	55.6	59.3	60.6	57.6	36.8	21.3	19.8	18.5
Jkstein	41.3	50.6	53.3	57.6	58.3	56.9	39.7	26.8	24.5	22.6
Saprol	40.6	43.3	46.3	49.6	52.6	57.9	45.8	36.6	32.3	29.8
Derosal	42.6	43.6	54.6	56.8	59.6	56.4	38.5	26.3	21.8	19.6
Topsin-M	41.3	46.6	50.3	53.6	55.3	57.8	43.6	32.3	30.7	27.6
Kitazin	40.6	45.3	45.6	48.3	50.6	57.2	46.8	38.7	35.7	32.7
Ronilan	41.3	44.6	46.3	49.3	51.3	57.3	41.5	37.8	35.3	32.8
Agrozim	42.6	49.3	52.6	54.3	56.6	56.8	40.3	30.4	28.9	25.8
Control	40.6	40.6	40.6	40.6	40.6	59.5	59.5	59.5	59.5	59.5

Table 24. Effect of fungicides and their concentrations on growth and sporulation of red rot pathogen

Treatment	Percentage											
	Redial growth (mm)				Dry mycelial mat (mg)				Sporulation (100 ml)			
	250 ppm	500 ppm	750 ppm	1000 ppm	250 ppm	500 ppm	750 ppm	100 ppm	250 ppm	500 ppm	750 ppm	1000 ppm
Bavistin	64.06	78.10	95.93	100.00	68.56	85.83	91.46	100.00	78.16	86.70	95.83	100.00
Jkstein	59.63	73.63	94.73	100.00	45.83	73.84	98.50	100.00	56.83	78.75	98.71	100.00
Saprol	30.36	62.84	85.80	100.00	36.43	68.64	89.66	100.00	49.83	84.83	100.00	100.00
Derosal	51.56	74.63	92.93	100.00	56.75	75.60	95.63	100.00	65.63	81.56	100.00	100.00
Topsin-M	37.46	63.94	94.18	100.00	41.80	70.63	95.63	100.00	40.82	86.72	100.00	100.00
Kitazin	37.96	51.56	79.93	94.16	31.63	59.66	86.70	94.00	43.36	65.73	91.83	92.00
Ronilan	33.26	65.19	83.56	96.00	40.60	73.46	90.78	96.30	44.63	.36	90.00	95.00
Agrozim	51.63	80.16	89.83	100.00	56.83	76.36	90.76	100.00	60.16	78.18	100.00	100.00

Table 25. Management of red rot disease through sett treatment with bioagents

Treatment		Germination (%)	Increase Over control	Mean	Settling mortality (%)	Reduction over control	Mean	Disease development (cm)	Reduction over control	Mean
T_1	Inoculated setts trated with Trichoderma viride for zero minute	17.6	7.98	16.26	16.5	28.6	30.43	23.5	27.69	30.77
T_2	Inoculated setts treated with T. viride for 15 minutes	20.3	24.54		15.5	32.61		21.5	33.85	
T_3	Inoculated setts treated with T. harzianum for zero minutes	18.0	10.43	21.47	14.0	39.13	42.39	22.0	32.31	34.61
T_4	Inoculated setts treated with T. harzianum for 15 minutes	21.6	32.52		12.5	45.65		20.5	36.92	
T_5	Inoculated setts treated with bavistin for zero minutes	22.3	36.81	42.94	15.0	34.78	41.30	18.5	43.08	45.38
T_6	Inoculated setts treated with bavistin for 15 minutes	24.3	49.08		12.0	47.83		17.0	47.69	
T_7	Inoculated setts (control)	16.3	-	-	23.0	-	-	32.5	-	-

The degradation of chlorophyll “a” and chlorophyll “b” contents in smutted cane leaves were recorded which suggested impaired synthesis of chlorophyll due to disturbed metabolic process in smutted cane plants. Sugarcane smut had also a marked effect on juice quality (Table- 26 & 27). In the juice of smutted canes, pH was low while titrateable acidity was fairly high. This may be attributable to smut pathogen (*Ustilago scitaminea,* Sydous) induced alteration resulting in the increased concentration of acids. In the juice of smutted cane, reducing sugars were accumulated while sucrose, brix, purity, commercial cane sugar, moisture, total soluble salts and ash were depleted. The depletion in sucrose and accumulation of reducing sugars suggested the inversion of sucrose induced by smut pathogen. Fairly high amount of glucose was accumulated in smutted cane juice indicating an increased invertase activity. Smut infection also disturbed the nutrient status in cane juice (Table 28 & 29).

The phosphorus, potassium, copper, manganese and zinc had depleted while nitrogen, iron, sodium, calcium, boron and magnesium had accumulated. Increase or decrease in these parameters varied according to the level of resistant of cane varieties (Table 30 & 31).

Table 26. Change in yield attributes due to smut infection

Yield attributes	Varieties			
	CoJ 64		BO 110	
	Healthily	Diseased	Healthily	Diseased
Length of cane (cm)	175.80	127.0	157.2	94.8
Girth of cane (cm)	8.3	5.7	6.4	4.2
Weight of cane (kg)	9.0	6.2	5.2	2.9
Moisture of cane (%)	76.3	71.7	77.0	69.2
Number of internode	18.5	13.4	16.1	11.1
Length of leaf (cm)	120.7	85.4	84.1	59.1
Width of leaf (cm)	4.0	2.6	3.2	1.8
Chlorophyll “a” (mg)/g leaf	0.129	0.067	0.188	0.102
Chlorophyll “b” (mg)/g leaf	0.138	0.073	0.204	0.121
Weight of juice (kg)	4.2	2.4	2.3	0.9

Table 27. Influence of smut on Cane Biometrix & Cholorophyll Contents

Yield attributes	Varieties			
	Co 1158		CoS 767	
	Healthily	Diseased	Healthily	Diseased
Length of cane (mm)	149.4	90.2	202.2	144.4
Girth of cane (cm)	5.6	3.7	7.3	4.2
Weight of cane (kg)	6.5	3.5	8.3	5.0
Moisture of cane (%)	74.7	68.2	79.0	70.0
Number of internode	17.1	11.1	22.7	14.7
Length of leaf (cm)	91.4	62.7	118.7	81.6
Width of leaf (cm)	2.3	1.4	3.5	2.1
Chlorophyll “a” (mg)/g leaf	0.276	0.069	0.216	0.098
Chlorophyll “b” (mg)/g leaf	0.296	0.098	0.227	0.103
Weight of juice (kg)	2.5	1.2	3.2	1.7

Table 28. Change in juice quality due to smut infection

Yield attributes	Varieties			
	CoJ 64		BO 110	
	Healthily	Diseased	Healthily	Diseased
Sucrose (%)	16.8	15.7	14.4	12.3
Brix (%)	19.4	18.6	17.7	17.1
Purity (%)	87.0	84.8	84.1	78.4
C.C.S (%)	10.9	10.0	8.90	7.9
pH	5.4	5.1	5.50	5.1
Tiraable acidity (m.e/litre)	0.318	0.0371	0.316	0.381
Ash (g/litre)	70.2	4.4	6.4	5.3
Reducing sugars (g/litre)	4.3	8.9	5.1	12.2
Soluble salts (g/litre)	50.8	45.6	43.6	40.1
Glucose (g/litre)	32.3	87.2	39.2	65.7
Moisture (%)	88.6	87.3	88.7	86.4

Table 29. Influence of smut on juice quality of cane

Juice constituents	Varieties			
	Co 1158		CoS 767	
	Healthily	Diseased	Healthily	Diseased
Sucrose (%)	15.7	14.8	16.2	15.2
Brix (%)	18.4	17.3	19.1	18.2
Purity (%)	86.8	81.3	85.0	83.6
C.C.S (%)	10.1	9.5	10.3	9.6
pH	5.6	5.0	5.5	5.1
Titratable Acidity (mea/litre)	0.225	0.307	0.227	0.285
Ash (g/litre)	6.2	4.5	5.8	3.3
Reducing sugars (g/litre)	2.8	4.8	3.4	6.8
Soluble salts (g/litre)	43.6	38.4	53.6	43.6
Glucose (g/litre)	35.1	83.2	31.0	61.4
Moisture (%)	86.0	83.1	87.4	86.4

Kumar et al.[20] further observed that type of smut infection also had adverse effect on sett germinability, growth parameters and juice quality (Table 32 & 33).

The incidence of smut was 88.6% in Co 1158 and 76.8% in BO 128 when setts from whip bearing cane were used as planting material, whereas, it was only 59.8% in Co 1158 and 51.6% in BO 128 with apparently healthy setts taken from smutted clumps. Whip bearing as well as apparently healthy canes gave poor germination. Reduction in thickness, length and weight of cane and juice was also observed both in whip bearing and apparently healthy canes. However, the extent of reduction was higher in first one. Reduction in the percentage of brix, polarity, purity and commercial cane sugar of whip bearing cane was observed as compared to healthy cane. Apparently healthy cane also showed reduction in these characters but in lesser extent.

The contents of total phenols, reducing sugars and amino acid in whip bearing and apparently healthy canes were fairly high in comparison to healthy one. However, the extent of their increase was higher in whip bearing cane as compared to apparently healthy cane of diseased stools. The increase in reducing sugars and decrease in non-reducing sugars both in whip bearing and apparently healthy cane may be attributed to the increased invertase enzyme activity. The concentration of ascorbic acid in smutted plants was lower as compared to healthy cane.

Table 30. Mineral nutrient status in smutted cane juice

Yield attributes	Varieties			
	CoJ 64		BO 110	
	Healthy	Diseased	Healthy	Diseased
Nitrogen	406.00	721.00	329.00	581.00
Calcium	150.00	200.00	220.00	350.00
Magnesium	250.00	260.00	265.00	300.00
Sodium	32.66	40.94	27.60	34.04
Boron	0.12	0.19	0.11	0.18
Iron	16.66	26.24	17.07	27.70
Phosphours	35.00	31.50	45.00	42.50
Potassium	58.25	32.97	55.96	30.65
Copper	1.54	1.47	1.59	1.45
Manganese	1.78	1.07	1.97	1.33
Zinc	37.32	22.57	38.48	27.23

Table 31. Mineral nutrient status in smutted cane juice

Nutrients (mg/lt.)	Varieties			
	Co 1158		CoS 767	
	Healthy	Diseased	Healthy	Diseased
Nitrogen	392.00	637.00	371.00	609.00
Calcium	200.00	300.00	260.00	400.00
Magnesium	240.00	270.00	210.00	230.00
Sodium	32.66	40.94	27.60	34.04
Boron	0052	0.103	0.078	0.151
Iron	18.74	27.70	19.740	30.320
Phosphours	35.00	32.50	36.500	35.500
Potassium	54.26	29.160	57.450	32.130
Copper	1.269	1.175	1.153	1.069
Manganese	2.00	1.300	1.850	1.140
Zinc	41.660	28.070	45.810	33.370

MANAGEMENT

Wind disseminated smut spores contaminate the exposed buds which may remain dorment or may cause dormant infections. Such diseased canes look apparently healthy. This situation is generally met with when smutted clumps grow in the vicinity of healthy clumps, in such cases, sett treatment with 0.25% Bayleton was most effective in eradicating external sett born infection, but was found ineffective in controlling natural infection.

WILT

Sugarcane wilt remained confined in south Bihar for a long period and from there, it spread slowly to other parts of Bihar due to unrestricted movement of planting materials. Many excellent, genotypes were withdrawn from the cultivation due to wilt problem.

ECONOMIC REPERCUSSIONS

Kumar and Kumar [21] observed that wilt disease of sugarcane caused by *Fusarium moniliforme* var. subglutinans produced adverse effect on yield attributes, deteriorated juice quality and disturbed nutrient status in cane juice (Table 34-36). However, the effect seems to be somewhat directly proportional to the level of resistance of cane varieties. The disease caused a reduction in sett germinability (40.2-50.1%), number of tillers (40.0-49.0%), millable cane (39.9-50.9%) and cane yield (45.2-51.2%), (Table 33) and juice content (42.8-59.2%), moisture content in cane (49.6-58.6%), leaf (41.3-51.4%) and juice (10.0-14.9%), Protein content in leaf (19.9-23.9%) and juice (18.8-34.5%), total chlorophyll (28.7-40.2%), pol (40.2-59.0%), brix (31.4-44.8%), purity (12.9-25.7%) and commercial cane sugar (44.5-66.0%). However, contents of reducing sugars and gum were increased in cane juice by 35.7-60.3 and 18.5-21.4 per cent, respectively, due to wilt infection.

Dohare et al. [22] reported that the content of phosphorus, potassium, copper, manganese and zinc seems to have been withdrawn, while nitrogen, iron, sodium, calcium and magnesium have been accumulated in the juice of wilted canes (Table 35).

The extent of reduction/accumulation on nutrients varied according to the level of resistance of cane varieties. Maximum reduction of Phosphorus (6.7%) was recorded in variety Co 1148 followed by 5.7% in CoJ 64, 4.5% in CoS 767 and 3.2% in BO 110. Potassium content in wilted cane juice showed 5.2 to 8.5% reduction.

The concentration of copper was also low which ranged from 4.6 to 7.4 per cent. There was a deletion of manganese content in wilted cane juice varying from 32.5 to 39.8 per cent depending upon cane varieties. Maximum reduction (39.8%) was found in Co 1148 followed by 38.4% in CoS 767, 35.0%. in CoJ 64 and 32.4% in BO 110. There was withdrawal of zinc content due to wilt infection ranging from 27.3 to 39.6 per cent. Maximum reduction (39.6%) was in Co 1148 whereas, it was 32.7% in CoJ 64 and 27.3% in both CoS 767 and BO 110.

Likewise, the contents of nitrogen, magnesium, iron, calcium and sodium were high in wilt infected cane juice in comparison to juice obtained from healthy canes. The accumulation of nitrogen in the wilted cane juice ranged from 62.5 to 76.2 per cent depending upon sugarcane varieties. Maximum accumulation (76.2%) was recorded in CoJ 64 followed by 73.4% in Co 1148, 63.5% in CoS 767 and 62.5% in BO 110. Wilt infected cane juice also contained high amount of magnesium. Maximum increase (15.4%) was observed in Co 1148 followed by 12.5% in CoJ 64, 9.5% in CoS 767 and 4.0% in BO 110. The content of iron was also increased in wilt infected cane juice ranging from 47.8 to 62.3 percent. Maximum increase (62.3%) was recorded in Co 1148 whereas, it was 61.8, 57.5 and 47.8 percent in CoJ 64, CoS 767 and Bo 110, respectively. The increase in the concentration of calcium varied from 5.2 to 8.4 per cent. It was maximum in Co 1148 (8.4%) followed by CoJ 64 (8.3%), CoS 767 (6.3%) and BO 110 (5.2%). The concentration of sodium in wilted cane juice was also increased varying from 22.6 to 25.9 per cent. The maximum increase was 2.59% in Co 1148, 24.5% in CoJ 64, 23.3% in BO 110 and 22.6% in CoS 767.

Management

Kumar et al. [23] observed that management of sugarcane wilt disease is almost impossible even with the most potent fungitoxicant. However, soil amendment with organic oil cakes reduced the propagules population of wilt pathogen in soil of sugarcane (*Fusarium moniliforme* var. subglutinans) and disease incidence (Table 37 & 38).

Mustard, groundnut, sesamum and castor cakes reduced soil population of wilt fungus significantly as compared to control. These cakes at 0.25% (w/w) and above caused significantly reduction in the number of propagules of wilt pathogen within 20 days. The suppressive effect was increased with a increase in the amount of oil cakes added to soil and with period of decomposition. After 60 days, there was significant reduction in fungal propagules at 2% concentration. Mustard and groundnut cakes, in general at 2% concentration were most effective and reduced the inoculums density of wilt pathogen by more than 65% over initial count. Sesamum and castor cakes were also effective but a higher concentration.

Soil amendment with these oil cakes also considerably reduced the incidence of sugarcane wilt disease. Maximum reduction in disease index (60.4%) was observed when soil was amended with groundnut cake followed by 59.74% with mustard cake, 48.8% with sesamum cake and 44.5% with castor cake. In general, the reduction in disease index was maximum at 2% concentration and minimum at 0.25%. Since, groundnut and mustard cakes at 2% level have effectively reduced the pathogen population, disease incidence and improved the plant growth thus cakes can be considered as possible amendment for the management of sugarcane wilt disease.

Survey and Surveillance

To know the disease position and varietal susceptibility, extensive surveys are being made since more than 25 years in the months of June, September and December in different parts of Bihar.

More than hundred sugarcane varieties were found affected with red rot, smut, wilt, red stripe, spike, top rot, mosaic, grassy shoot, sett rot, ratoon stunting, leaf scald & leaf spot diseases singly or in combination on a particular variety. The severity of the diseases was comparatively higher in the months of September and December as compared to the month of June. However, the severity of the major diseases particularly red rot, smut and wilt diseases is under manageable dimension due to deployment of anti red rot staff in reserved areas of different sugar factories (Table 39-40).

Table 32. Effect of smut infection on yield attributes of sugarcane

Type of cane samples	Variety	No. of cane evaluated	Germination (%)	Smut incidence (%)	Wt. of cane (kg)	Wt. of juice (kg)	Extraction (%)	Cane height (cm)	Cane girth (cm)
Whip bearing cane from diseased clumpsz	BO 128	25	16.8	76.8	16.7	8.9	53.3	76.5	4.3
	Co 1158	25	13.4	88.6	12.3	5.6	45.5	71.5	4.1
Apparently healthy cane from diseased clumps	BO 128	25	30.8	51.6	31.6	12.4	57.4	130.6	4.9
	Co 1158	25	27.6	59.8	25.7	8.4	53.5	105.4	4.7
Healthy cane from healthy clumps (control)	BO 128	25	50.6	0.0	38.1	27.8	73.6	208.3	6.2
	Co 1158	25	46.0	0.0	30.5	19.2	67.1	193.6	5.7

Table 33. Effect of smut infection on juice quality of sugarcane

Type of cane samples	Variety	Brix (%)	Pol (%)	Purity (%)	CCS (%)	Total phenols (mg/g of dry tissue)	Reducing sugars (mg/g of dry tissue)	Non-reducing sugar (mg/g of dry tissue)	Free amino acid (mg/g of dry tissue)	Ascorbic acid (mg/g of dry tissue)
Whip bearing cane from diseased clumps	BO 128	14.6	10.5	71.9	6.5	1.57	1.61	14.7	0.65	63.4
	Co 1158	14.0	10.1	72.1	6.2	1.51	1.57	13.3	0.59	67.6
Apparently healthy cane from diseased clumps	BO 128	16.9	13.6	80.5	8.2	1.43	1.31	15.8	0.46	67.7
	Co 1158	15.9	12.3	77.3	7.9	1.39	1.17	14.7	0.41	70.6
Healthy cane from healthy clumps (control)	BO 128	19.0	15.8	83.2	10.6	0.93	1.03	16.7	0.28	78.7
	Co 1158	18.6	14.9	80.1	9.8	0.86	1.10	15.9	0.21	76.3

Table 34. Effect of wilt disease on quantitative parameters of sugarcane

Parameters	Varieties											
	BO 113			BO 120			BO 125			BO 139		
	Unioculated cane	Inoculated cane	Decrease (%)	Unioculated cane	Inoculated cane	Decrease (%)	Unioculated cane	Inoculated cane	Decrease (%)	Unioculated cane	Inoculated cane	Decrease (%)
Disease Index (0-4)	-	2.8	-	-	1.4	-	-	2.4	-	-	1.9	-
Germination (%)	45.3	22.6	50.1	46.5	27.8	40.2	44.4	24.2	45.5	43.9	24.8	43.5
No. tillers ('000 ha)	165.2	82.9	49.8	167.8	100.6	40.0	163.8	90.9	44.5	162.6	94.7	41.7
No. millable canes ('000 ha)	92.6	46.3	50.0	94.8	57.0	39.9	90.9	49.7	45.3	90.6	52.4	42.2
Yield (t/ha)	64.8	31.6	51.2	66.4	36.4	45.2	63.6	33.4	47.5	63.3	34.8	45.0
Weight of juice (kg)	3.2	1.3	59.3	4.9	2.8	42.8	3.8	1.8	52.6	4.6	2.3	50.0

Table 35. Effect of wilt infection on mineral nutrient contents of cane juice (mg/l juice)

Para-meters	Varieties											
	Co 1148			CoJ 64			CoS 767			BO 110		
	Heal-thy	Diseased	Increased/ Decreased (%)	Healthy	Diseased	Increased/ Decreased (%)	Healthy	Diseased	Increased/ Decreased (%)	Healthy	Diseased	Increased/ Decreased (%)
Disease index (0-4)	-	3.1	-	-	2.9	-	-	2.3	-	-	2.1	-
Nitrogen	334.0	579.0	73.4	408.0	719.0	76.2	375.0	613.0	63.5	395.0	642.0	62.5
Phosphorus	279.5	260.5	6.7	280.3	264.0	5.7	286.4	273.6	4.5	284.2	275.2	3.2
Magnesium	260.0	300.0	15.4	240.3	270.0	12.5	210.0	230.0	9.5	250.0	260.0	40.0
manganese	1.78	1.07	39.88	2.0	1.3	35.0	1.8	1.1	38.4	1.9	1.3	32.5
Zinc	3.73	2.25	39.67	4.2	2.8	32.7	3.7	2.7	27.3	4.6	3.3	27.3
Copper	1.26	1.18	7.64	1.2	1.1	7.3	1.538	1.467	4.6	1.538	1.453	5.5
Iron	17.07	27.70	62.3	18.7	30.3	61.8	16.7	26.2	57.5	18.7	27.7	47.8
Potash	1180.0	1080.0	8.47	1140.0	1045.0	8.3	1120.0	1050.0	6.3	1160.0	1100.0	5.2
Calcium	230.0	360.0	56.5	160.0	210.0	31.3	270.0	410.0	51.9	210.0	310.0	47.6
Sodium	35.50	44.7	25.9	33.6	41.9	24.5	42.7	52.4	22.6	27.6	34.0	23.3

Table 36. Effect of wilt disease on qualitative parameters of sugarcane

Parameters	Varieties											
	BO 113			BO 120			BO 125			BO 139		
	Healthy	Diseased	Increase/ Decrease %	Healthy	Diseased	Increase/ Decrease %	Healthy	Diseased	Increase/ Decrease %	Healthy	Diseased	Increase/ Decrease %
Moisture in cane (%)	65.15	27.11	58.38	64.05	32.28	49.60	60.13	28.24	53.03	62.22	30.37	51.18
Moisture in cane leaf (%)	70.89	34.35	51.54	75.72	44.48	41.25	74.53	37.42	49.79	72.25	39.72	45.02
Moisture in cane juice (%)	80.25	92.21	14.90	79.92	87.92	10.01	81.05	92.19	13.75	80.73	90.62	12.25
Protein in leaf (%)	7.78	5.92	23.90	11.83	9.47	19.94	10.99	8.46	23.02	8.28	6.42	22.46
Protein in juice (%)	1.10	0.72	34.54	1.12	0.91	18.75	0.83	0.59	28.91	1.17	0.89	23.93
Chlorophyll "a" (mg/g. f.w.)	0.48	0.31	35.41	0.53	0.36	32.07	0.59	0.34	42.37	0.62	0.42	32.23
Chlorophyll "b" (mg/g. f.w.)	0.49	0.27	44.89	0.41	0.31	24.39	0.48	0.33	31.25	0.58	0.39	32.75
Total chlorophyll "(a + b") (mg/g.f.w.)	0.97	0.58	40.20	0.94	0.67	28.72	1.07	0.67	37.38	1.20	0.81	32.50
Pol %	16.88	6.92	59.00	15.56	9.30	40.23	17.47	8.60	50.77	16.53	9.16	44.58
Brix (%)	19.40	10.71	44.80	17.60	12.08	31.35	19.60	11.60	40.82	18.80	12.21	35.05
Purity (%)	87.01	64.61	25.74	88.40	76.98	12.92	89.13	74.14	16.82	87.92	75.02	14.67
Reducing sugar (mg/100 ml. Juice)	0.58	0.93	60.34	0.70	0.95	35.71	0.52	0.79	51.92	0.56	0.83	48.21
CCS %	11.90	4.05	65.97	11.06	6.14	44.48	12.46	5.55	55.45	11.72	5.95	49.23
Gum	0.28	0.34	21.42	0.27	0.32	18.51	0.51	0.37	19.35	0.26	0.31	19.23
Soluble salt (g/lt.)	31.75	47.01	48.08	32.03	45.82	43.06	31.27	46.02	47.17	31.01	45.44	46.53
pH	5.54	5.47	1.26	5.23	5.20	0.57	5.24	5.19	0.95	5.28	5.24	0.75
T. Acidity (Meg/lt.)	0.32	0.44	37.50	0.34	0.43	26.47	0.33	0.44	33.33	0.33	0.43	30.30

Table 37. Effect of different oil cakes on population of (*Fusarium moniliforme*) in soil

Oil cake	Number of propagules (10 3 g-1) at days after mixing the oil cakes				
	Concentration (w/w)	20	40	60	Mean
Castor	0.25 0.50 1.00	17.0 16.7 14.0 21.3	15.7 12.3 10.0 15.3	14.7 10.7 7.3 7.3	13.5
Sesamum	0.25 0.50 1.00 2.00	14.3 12.7 10.7 9.3	14.0 12.7 9.3 8.7	13.3 10.0 6.7 6.7	10.7
Mustard	0.25 0.50 1.00 2.00	15.7 13.7 10.0 6.0	13.3 10.0 18.0 2.3	10.7 6.0 3.3 0.7	8.4
Groundnut	0.25 0.50 1.00 2.00	17.3 14.3 10.0 8.0	13.7 10.7 18.0 4.3	10.0 7.3 4.7 0.7	9.1
Control (inoculated)		26.3	24.3	27.7	26.1

Table 38. Effect of different cake amendments on development of wilt (*Fusarium moniliforme*) in sugarcane

Oil cake	Concentration (w/w)	Disease index	Mean of disease index	Reduction in disease index (%)	Mean of reduction in disease index
Castor	0.25 0.50 1.00 2.00	2.9 2.3 2.1 1.8	2.3	29.3 43.9 48.7 56.1	44.5
Sesamum	0.25 0.50 1.00 2.00	2.6 2.3 2.0 1.2	2.0	36.6 43.9 51.2 63.4	48.8
Mustard	0.25 0.50 1.00 2.00	2.0 1.9 1.5 1.2	1.6	51.2 53.6 63.4 70.7	59.7
Groundnut	0.25 0.50 1.00 2.00	2.3 1.9 1.4 0.9	1.6	43.9 53.6 65.9 78.6	60.4
Control (inoculated)		4.1			

Table 39. Survey report conducted at SRI, Pusa, Bihar during 2005 to 2009

Sl. No.	2005-06		2006-07		2007-08		2008-09	
	Varieties	Diseases	Varieties	Diseases	Varieties	Diseases	Varieties	Diseases
1.	BO 70, BO 120, BO 138, CoS 687, CoSe 93232, CoS 92423 & CoSe 98235	Red rot, Wilt	BO 128	Red rot, Wilt, Smut	Co 1158	Smut	CoSe 95422, CoS 8436, Co 239, CoLK 8102, BO 138, CoBln 05501 & CoS 1148.	Red rot & Wilt
2.	BO 110, BO 130, BO 133 & BO 147	Wilt	BO 120, Copant 84212, Co 0236, Co 229, CoS 687, CoS 95222, UP 0090, CoLK 8102, CoLK 9306, CoJ 83 & CoJ 84	Red rot, Wilt	Co 1148, CoLK 8102, CoS 767, BO 120, BO 128, BO 138, Co 0233, CoJ 64 & CoBln 94063	Red rot	BO 128	Red rot, Wilt, Top rot & Smut
3.	BO 128	Red rot, Wilt, Smut	BO 138, CoBln 92202, CoBln 03173, CoBln 03171, CoBln 03172, CoBln 94063 & CoSe 1424,	Wilt	CoBln 9006, CoBln 02173, CoBln 03171, CoBln 03172, CoBln 03173 & CoJ 83	Wilt	Co 05021	Red rot
4.	-	-	-	-	-	-	UP 9530 & CoP 02182	Wilt
5.	-	-	-	-	-	-	BO 70	Red stripe & Spike
6.	-	-	-	-	-	-	Co 1158	Wilt & Smut
7.	-	-	Co 233	Red rot, Wilt, Top rot	BO 74	Red stripe	-	-
8.	-	-	Co 0235 & CoBln 9006	Leaf spot	BO 128	Top rot	-	-
9.	-	-	CoSe 95423 & CoSe 0235	Smut, Wilt	-	-	-	-
10.	-	-	CoSe 98231	Smut	-	-	-	-

Table 40. Survey report conducted at SRI, Pusa, Bihar during 2009 to 2013

Sl. No.	2009-10		2010-11		2011-2012		2012-2013	
	Varieties	Diseases	Varieties	Diseases	Varieties	Diseases	Varieties	Diseases
1.	BO 70	Red rot, Red stripe & Spike	CoSe 95422, Co 0238, Co 98014, Co 239, Co 1148, CoSe 8436, BO 138, CoBln 05501 & CoB 99161	Red rot	CoSe 95422, Co 0238, Co 98014, Co 239, Co 1148, CoS 8436, BO 138, CoBln 05501 & CoB 99161	Red rot	BO 137 & BO 147	Smut, Wilt, Pokkah boeing
2.	BO 128	Smut, Red rot, Wilt & Top rot	CoB 99161, CoSe 95422, Co 0238, Co 98014, Co 0239, BO 128 & BO 138	Wilt	CoB 99161 & CoSe 95422	Wilt	BO 145, CoS 91269 & CoS 8436	Wilt, Red rot
3.	BO 138, Co 0238, Co 1148, Co 1158, CoS 8436, CoLK 8102, CoSe 95422, CoBln 05501	Red rot & Wilt	BO 128, , Co 05019 & CoBln 04174	Top rot	CoSe 98231 & BO 150	Smut	BO 141	Sett rot & Pokkah boeing
4.	-	-	Co 0118	Pine apple, Sett rot	Co 05019 & CoBln 0404	Top rot	Co 238	Smut & Pokkah boeing
5.	-	-	CoB 99161	Mosaic	CoB 99161	Mosaic	CoS 8432 & CoSe 98231	Smut
6.	-	-	-	-	-	-	CoJ 64	Spike
7.	-	-	-	-	-	-	CoLk 84184	Pokkah boeing
8.	-	-	-	-	-	-	CoSe 95422	Red rot, Wilt & Top rot
9.	-	-	-	-	-	-	Co 235	Wilt, Pokkah boeing & Red rot
10.	-	-	-	-	-	-	-	-

CONCLUSION

Blight, mortality and root rot are important fungal diseases of sugarcane seedlings causing considerable damage to the seedlings in seed bed nursery. The fungi and their filtrates caused seedling mortality and retarded the seed germinability. The diseases were adequately managed when fungicides were applied at the time of sowing or after five days of emergence. Pre-sowing soil treatment as well as seed treatment with Bavistin, Thiram, Captan and Brasbicol were equally effective against root rot disease.

Red rot, Smut and Wilt diseases retarded the yield attributes, juice quality and caused the imbalance in nutrients status in cane and its juice. Red rot infection also exerted adverse affect on seed germinability, increased the incidence of red rot and settling mortality. Red rot fungus in association with wilt fungi caused more damage than red rot fungus alone. With an increase in the level of red rot infection, the extent of losses in yield attributes and juice quality was also increased. Maximum loss was recorded 100% infestation level. Sett treatment with Bavistin (0.1%) was found to be more efficacious than bio-agents in enhancing the sett germinability and in reducing red rot development. External smut infection was eradicated when the setts were treated with Baylaton (0.25%) before planting. Soil amendment with organic cakes (2%) considerably reduced the wilt pathogen population in soil and thereby reduced the disease incidence and improved the plant growth.

REFERENCES

[1] Kumar S, Dwivedi NB, Singh RN and Mishra MM (1986). Seedling mortality disease of sugarcane in Bihar. *Bhartiya Sugar,* 11 (4): 45-49.

[2] Rai B, Kumar S and Kumar B (2002). Evaluation of fungicides against seedling blight disease in sugarcane. *Annals of Biology,* 18 (2): 143-146.

[3] Kumar S (1989). Control of sugarcane seedling root rot in seed bed nurseries. *RAU J. Res.,* 7 (1-2): 97-99.

[4] Kumar S, Dwivedi NB, Mishra MM and Sinha RN (1988).Variation in *Colletotrichum falcatum,* the incitant of red rot disease of sugarcane in Bihar. *Bhartiya Sugar,* 13 (3): 53-56.

[5] Kumar S (1995). Management of red rot disease of sugarcane with agrochemicals. Proc. National Seminar on Strategies for Research and Management of red rot pp. 331-338.

[6] Md. Minnatullah and Kumar S (2005). Pathogenic variability in Colletotrichum falcatum. *Ann. Pl. Protec. Sci.,* 13 (2): 465-529. 500-501.

[7] Kumar S, Singh NP, Kumar V and Dwivedi NB (1994). Deterioration in yield and juice quality parameters of sugarcane due to isolates of red rot pathogen. *Ind. J. Sugarcane Technology,* 9 (2): 115-122.

[8] Kumar S, Kumar B and Kumar V (2000). Changes in phenolic and amino acid contents of sugarcane induced by isolates of red rot pathogen. *Annals of Agri. Bio. Res.,* 5 (1): 75-78.

[9] Pandey SP, Kumar S, Dwivedi NB, Kumar V and Kumar B (2000). Losses in sugarcane varieties BO 70 and BO 74 due to red rot. *Ind. Phytopath,* 53 (3):333-334

[10] Kumar S (1992). Juice quality and nutrient status in red rot infected cane juice. *Bihar. Bhartiya Sugar,* 17(4): 23-30.

[11] Kumar S, Singh NP, Dwivedi NB and Kumar V (1996). Metabolic changes in sugarcane induced by Pathotypes o red rot pathogen. *Ind. J. Sugarcane Technology,* 11 (1): 78-81.

[12] Kumar S, Dwivedi NB and Kumar V (1997). Quantitative and qualitative losses in sugarcane due to isolates of red rot pathogen in Bihar. Proc. National Seminar on Sucrose Synthesis and recovery in Sugarcane: Issues and Dimensions. Pp. 145-148.

[13] Kumar S, Singh NP, Dwivedi NB and Kumar V (1999). Changes in chemical composition of cane juice due to pathotypes of red rot pathogen. *Bhartiya Sugar,* 24 (12): 9-13.

[14] Kumar U, Kumar S and Dohare S (2002). Losses in yield parameters and juice quality of sugarcane due to pathotypes of red rot pathogen. *Sci. and Cult.,* 68 (7-8): 205-206.

[15] Md. Minntullah, Kumar S and Thakur MB (2006). The effect of red rot pathotypes on nutrient status in cane juice. *Indian sugar;* 29-31.

[16] Md. Minntullah, Thakur MB and Kumar S (2007). Effect of Colletotrichum falcatum Pathotypes on yield attributes and juice quality of sugarcane. *Indian sugar,* 55-60.

[17] Md. Minnatullah, Kumar S, Akhtar R and Dohare S (2012). Management of red rot disease through sett treatment with biogents. *Environment and Ecology,* 28 (4A): 2660-2661, 2012.

[18] Kumar S, Sinha RN and Rai B (1991). Influence of micronutrient spray on red rot development, yield attributes and juice composition of sugarcane. *Sci and Cult.,* 57 (3-4): 83-86.

[19] Md. Minnatullah, Kumar S, Dohare S and Akhtar R (2010). Evaluation of sugarcane genotypes against red rot, wilt and smut diseases. *Indian sugar, July* 2010, 23-26.

[20] Kumar S, Rai B, Dohare S and Kumar V (2001). Losses in sugarcane and its juice due to smut infection in Bihar. *Sci. and Cult.,* 67 (5-6): 171-172.

[21] Kumar S and Kumar S (2005). Effect of wilt disease on quantitative and qualitative parameters of sugarcane Indian sugar; 23-27.

[22] Dohare S, Kumar S and Kumar B (2003). Deterioration in mineral nutrient status of sugarcane juice due to wilt pathogen. *Ann. of Agril. Bio. Res.,* 8 (2): 243-246.

[23] Kumar B. Kumar S and Dohare S (2003). Management of wilt disease of sugarcane with organic amendments. *Crop. Res.,* 25 (1): 205-208.

In: Current Status of Sugarcane Research in India
Editors: A. K. Tiwari, M. Lal and A. K. Singh
ISBN: 978-1-63463-458-8

Chapter 8

EVALUATION OF SUGARCANE GENOTYPES TO RED ROT DISEASE IN THE FLOOD PRONE TRACTS OF KERALA

A. Sajeena*, M. Surendran, V. R. Shajan, Beena Thomas, Bindhu. J. S, Jessy M. Kuriakose, Reena Mathew and Sosamma Cherian

Sugarcane Research Station, Kerala Agricultural University, Thiruvalla, India

ABSTRACT

During 2008 - 2012, 44 sugarcane genotypes were screened against red rot disease by the standard plug and nodal methods of artificial inoculation. Among them, only 8 genotypes showed moderately resistant (MR) reaction to both the methods of artificial inoculation. Nine sugarcane genotypes exhibited resistance (R) reaction against red rot disease and also high brix (>20%) which enable them to be released as a new variety or may be used as parents in hybridization programmes. Most of the sugarcane genotypes exhibited resistance reaction (R) to red rot by nodal method of artificial inoculation. 12 genotypes exhibited susceptibility (S) to nodal method of inoculation which deprives of their chance for further screening. 14 genotypes exhibited a variable reaction to red rot disease which may be attributed to the varying climatic factors prevalent during the period of study.

Keywords: Sugarcane, *Colletotrichum falcatum*, red rot, varietal screening

INTRODUCTION

Sugarcane is the source of 70% of the world's sugar. It has a very long history of cultivation in the Indian sub-continent of which the earliest reference is in the Atharva Veda

* Corresponding author Email: sajeenamanjima@gmail.com.

(1500-800 BC). The cultivation of sugarcane in Kerala is widely distributed among the amazingly varying and climatologically distinct tracts of the flood prone river banks of Central Travancore, semi arid tracts of Palakkad, and rain shadow, hill tract terrains of Idukki district. The soil of the river banks of the Central Travancore region is under flood prone situation and gets inundated during the South- West and North - East monsoons, rendering sugarcane to be the most suitable crop for the region.

Red rot is considered as a century old problem of sugarcane in India. It is an important disease of sugarcane responsible for considerable yield loss and also for the elimination of many commercial varieties in India. It is caused by a fungus *viz.*, *Colletotrichum falcatum* Went. The disease symptoms can be noticed during the cane formation stage as sudden yellowing of 3rd and 4th leaves. Later these leaves get gradually dried up. The presence of cross wise white spots alternated with red spots in the internal tissues of the stalk is the characteristic and diagnostic symptom of the disease. The pathogen affects sugarcane juice quality, brix, sucrose percentage, purity and CCS percentage [1]. The disease is mainly seed piece transmissible and largely systemic in nature, which makes its management through fungicides a difficult task [2]. The difficulty in managing sugarcane red rot disease through chemotherapy is due to the impervious nature of rinds and fibrous nodes at cut ends of the crop [3]. The use of resistant varieties is the only means of controlling red rot disease effectively. The resistant varieties which succumb to red rot disease as well as susceptible varieties are continuously being replaced with resistant or moderately resistant ones [4].

Sugarcane Research Station (SRS), Thiruvalla under Kerala Agricultural University succeeded in replacing the prominent, but highly red rot susceptible sugarcane variety, Co 997 with resistant varieties evolved from the trials of All India Co-ordinated Research Project on Sugarcane. The research efforts of the station led to the evolution and release of a series of high yielding, high sugared varieties *viz.,* Madhuri, Madhumathy, Madhurima and Thirumadhuram. These varieties are red rot resistant and are also suited to the flood prone and semi arid tracts of Kerala. However, the variability and evolution of new virulent races of the red rot pathogen may cause breakdown of resistance in resistant varieties and is a recurring problem in red rot management [5]. According to Narendra Singh et al. [6], favorable climatic conditions as well as drought and flooding of the host plants are the most conducive factors for a red rot epidemic. Hence there is a need for continuous screening of sugarcane varieties to red rot disease. The present study was taken up to screen sugarcane genotypes against red rot disease under flooded conditions of Kerala during a period of five years (2008-12).

Materials and Methods

The experiment was conducted at SRS, Thiruvalla during 2008-2012 for two seasons under each trial. A total number of 44 sugarcane genotypes were screened for their reaction to red rot disease. The disease screening in sugarcane genotypes was done by adopting the standardized plug [7] and nodal methods of artificial inoculation. By plug method, inoculation was done in the middle of the 3rd exposed internode from bottom and two drops of the spore suspension of the fungus was injected in each cane and sealed with plastic clay or modeling clay. Freshly sporulating 7-day-old culture of the fungus was used for the inoculation. The isolate of red rot pathogen used for the inoculation was the designated pathotype (CF06) for the peninsular zone. By nodal method, the inoculation was done by pouring 1 ml of fungal spore suspension into the axils of the 4th and 5th nodes from the top after slightly pulling the leaf sheath between the sheath and the stem of two opposite buds [8]. One six-meter row of 20 sugarcane clumps was used for artificial inoculation of the red rot pathogen. Five red rot susceptible varieties of the same maturity group were used as the standards. Inoculation was done with the onset of pre-monsoon under high humid conditions. Disease scoring was done two months after the inoculation. Inoculated canes free from borer infestation and other damages were selected for symptom observation. The canes were split opened longitudinally along the point of inoculation for observation of the disease symptoms. These were graded on the international scale of 0-9 [9] as follows:

1. Condition of the top, if Green (G) = O, Yellow (Y)/Dry (D) =1,
2. Lesion width examined in the internodes above the inoculated internode were assigned the score 1, 2 or 3,
3. White spots were assigned score of 1 or 2 according to whether it is restricted or progressive,
4. Nodal transgression, which is the number of nodes crossed above the inoculated internodes or scores and were assigned as (1) - if one node crossed, (2) - if two nodes crossed and (3) if three or more crossed (maximum).
    - Average Score = Total Score/No. of canes evaluated.
    - The genotypes were categories as

0.0 to 2 - R,
2.1 to 4 – MR,
4.1 to 6 – MS,
6.1 to 8 – S and
Above 8 – HS

RESULTS AND DISCUSSION

Screening of Sugarcane Genotypes to Red Rot by Plug & Nodal Methods of Inoculation

In the present study, among the 16 sugarcane genotypes screened against red rot, 4 genotypes *viz*., Co Snk 03754, Co 0403, Co 0409 and Co 94008 showed resistant to moderately resistant reaction (R/MR) to both the plug and nodal methods of inoculation, whereas 8 genotypes *viz*., Co Snk 03632, Co Snk 03822, Co 7219, CoM 0326, CoM 0316, Co 0416, Co 86032 and CoC 671 showed susceptible to moderately susceptible reaction (S/MS) to plug method for both the seasons (2008-09 & 2009-10) tested (Table 1). During 2009-10 & 2010-11, out of the 17 genotypes tested, only two genotypes, *viz*., Co Snk 05104 and Co VSI 05122 showed resistance (R) reaction to both methods of inoculation, whereas 10 genotypes *viz*., Co 05002, Co Snk 05101, Co Snk 05103, CoM 05082, Co VSI 05123, Co 86032, Co 7219, Co 94008, Co 85004 and CoC 671 exhibited susceptible (S) reaction to red rot disease (Table 2).

Among the 11 genotypes tested during 2010-11 & 2011-12, only two genotypes *viz*., Co 07012 and Co 07015 showed resistance reaction whereas 5 genotypes *viz*., Co 07007, Co 07008, Co 07009, Co 07010 and CoC 671 exhibited susceptible reaction to red rot (Table 3). Khan et al [10] reported similar results while screening 40 sugarcane varieties for the source of resistance against red rot disease in a field trial. 20 varieties displayed resistance reaction, while 7 varieties exhibited moderately resistance reaction. The remaining test varieties exhibited moderately susceptible to highly susceptible reaction against the red rot disease. In an experiment conducted during 2004-05 to 2005-06, out of the 28 promising sugarcane varieties tested, 3 varieties possessed relative resistance reaction, 7 varieties showed moderately resistance reaction, 8 varieties showed moderately susceptible reaction, 8 varieties exhibited susceptible reaction and 2 varieties showed highly susceptible reaction to red rot disease [11].

These moderately resistant (MR) and resistant (R) varieties may be utilized in breeding programmes aimed at red rot disease resistance. For successful exploitation of sugarcane genetic resources for disease resistance and better yield attributes, it is very necessary to carefully characterize and evaluate different sugarcane varieties among different species. *Saccharum spontaneum* parents have been proved to be obviously appropriate for hybridization programmes to transfer a single specific character e.g. disease resistance to other sugarcane genotypes with good yield and quality attributes.

Table 1. Evaluation of sugarcane genotypes during 2008 – 09 and 2009 – 10

	2008- 2009				2009-10		
Si. No	Varieties	Red rot reaction		Brix percentage	Red rot reaction		Brix percentage
		Plug Method	Nodal Method		Plug Method	Nodal Method	
1	Co Snk 03632	S	R	18.5	S	R	18.7
2	Co Snk 03754	R	R	19.0	MR	R	20.0
3	Co Snk 03822	MS	R	16.0	S	S	18.8
4	MS 0301	S	R	17.7	R	R	19.2
5	Co M 0326	MS	R	17.7	S	R	17.3
6	Co 0403	MR	R	20.3	MR	R	20.2
7	Co N 03131	MS	R	19.5	MR	R	17.4
8	Co 0415	MR	R	16.7	MS	R	19.6
9	Co 0409	R	R	16.3	MR	R	17.8
10	Co M 0316	S	R	16.0	S	S	17.7
11	Co 0416	MS	R	16.3	S	S	18.7
	Standards						
12	Co 7219	S	R	16.0	S	R	18.8
13	Co 86032	S	R	18.7	MS	S	19.5
14	Co 94008	MR	R	18.8	R	R	16.7
15	Co 85004	MR	R	20.0	MS	R	19.9
16	CoC 671	S	R	21.2	HS	S	20.8

Table 2. Evaluation of sugarcane genotypes during 2009 – 10 and 2010 – 11

	2009-10				2010-11		
Si. No	Varieties	Red rot reaction		Brix percentage	Red rot reaction		Brix percentage
		Plug method	Nodal method		Plug method	Nodal method	
1	Co 05001	R	R	18.4	MS	R	18.3
2	Co 05002	MS	R	17.8	S	S	18.7
3	Co 05007	MR	R	19.4	MS	R	18.6
4	CoN 05071	MR	R	17.9	MS	S	17.7
5	Co Snk 05101	MS	R	17.8	MS	R	18.0
6	Co Snk 05103	S	S	16.8	MS	R	17.7
7	Co Snk 05104	R	R	20.3	MR	R	20.3
8	Co Snk 05105	MS	R	19.1	MR	R	14.8
9	CoM 05082	S	S	17.6	MS	R	18.0
10	Co VSI 05121	MR	R	19.4	MS	R	19.7
11	Co VSI 05122	R	R	18.6	MR	R	19.0
12	Co VSI 05123	MS	R	18.7	MS	R	17.2
	Standards						
13	Co 86032	MS	R	20.6	MS	R	20.5
14	Co 7219	MS	R	19.9	MS	R	18.7
15	Co 94008	MS	R	16.6	MS	R	17.8
16	Co 85004	MS	R	19.1	MS	R	18.9
17	CoC 671	S	S	20.3	MS	R	19.5

Table 3. Evaluation of sugarcane genotypes during 2010 – 11 and 2011– 12

	2010-11				2011-12		
Si. No	Varieties	Red rot reaction		Brix percentage	Red rot reaction		Brix percentage
		Plug method	Nodal method		Plug method	Nodal method	
1	Co 07006	MR	R	18.3	MS	R	18.9
2	Co 07007	MS	R	18.3	S	R	20.3
3	Co 07008	MS	R	19.5	MS	S	19.2
4	Co 07009	MS	R	21.3	MS	R	17.8
5	Co 07010	MS	R	18.7	MS	R	16.9
6	Co 07012	MR	R	15.8	MR	R	20.8
7	Co 07015	MR	R	17.2	MR	R	21.8
8	Co N 07071	MS	R	17.8	MR	R	18.2
9	PI 07131	MS	R	17.5	MR	R	20.5
	Standards						
10	Co 85004	MR	R	19.0	S	R	19.6
11	Coc 671	MS	R	19.6	S	S	21.8

Resistance and Brix Percentage of the Sugarcane Genotypes

Among the 44 sugarcane genotypes tested, 12 genotypes including 3 standards (Co 86032, Co 85004 & Coc 671) exhibited high brix % (Table 1, 2 & 3).

Two sugarcane genotypes exhibited high brix % along with resistance reaction to red rot disease for both the years tested whereas 7 genotypes showed high brix % as well as resistance to red rot for at least one year tested.

Malathi and Viswanathan [12] correlated host resistance and sucrose content of various sugarcane genotypes with pathogenic virulence. Cultural studies indicated that *C. falcatum* virulence related factors *viz.*, growth, sporulation and conidial germination had negative correlation with host resistance and positive correlation with sucrose content of various sugarcane cultivars. They observed that sucrose content played a major role in deciding the host resistance.

Reaction of Genotypes to Nodal Method of Inoculation

For the entire trials carried out during the period from 2008-09 to 2011-12, most of the genotypes were found to exhibit resistance (R) reaction to red rot by nodal method of artificial inoculation.

Among the 44 genotypes screened for their reaction to red rot, 12 genotypes exhibited susceptible reaction by nodal method of inoculation (Table1, 2 & 3). Nodal method is a comparatively less severe method of artificial inoculation compared to plug method where no artificial wounds are induced for inoculation of the red rot pathogen. In plug method of artificial inoculation, intentional wounds are introduced in the internodes of the crop and red rot pathogen will be inoculated in the wounds for symptom development and thus varietal screening will be done.

Hence the varieties exhibiting susceptible reaction to nodal method of artificial inoculation will not be considered for further breeding programmes. The breakdown of red rot resistance is primarily attributed to the appearance of new strains/pathotypes of red rot pathogen. Rahman [13] reported that light types of red rot pathogen isolates are generally more virulent than other types of the pathogen.

Variation in Disease Reaction among Genotypes and Environmental Factors

Some of the sugarcane varieties could exhibit the same reaction for both the years under trial. Some of the sugarcane genotypes tested showed variation in their disease reaction during the period of study.

This may be due to the change in the environmental factors which were prevailing during the year of study (Table 4). Different **environmental factors** (temperature, rainfall, soil moisture) may also be responsible for enabling infection of sugarcane genotypes by red rot pathogen [14].

Table 4. Weather parameters during the period of study

Month	2008			2009			2010			2011			2012		
	R.F	R.D	R.H	R.F	R.D	R.H	R.F	R.D	R.H	R.F	R.D	R.H	R.F	R.D	R.H
Jan	Nil	Nil	83.5	3	1	77.4	8	2	79.1	118	6	85.6	4	2	77.4
Feb	148	4	71.3	Nil	Nil	77.4	Nil	Nil	83.3	52	2	83.5	Nil	Nil	81.3
Mar	159	11	88.8	49	5	81.5	75	2	93.1	5	2	82.5	27	6	82.4
Apr	178	9	90.3	86	6	80.5	168	11	98	329	15	82.9	127	13	81.8
May	27	3	92.8	61	8	90.5	209	13	99.4	363	9	81.8	150	9	85.2
June	258	22	90.5	343	22	91.7	475	26	99.8	471	22	88.4	161	17	84.0
July	706	23	87.3	494	26	95.7	367	24	99.3	309	25	84.7	-	-	-
Aug	191	16	85.3	136	19	87.3	299	21	99.1	240	17	84.0	-	-	-
Sept	295	12	81.8	256	14	90.8	177	16	96.4	388	20	83.3	-	-	-
Oct	283	18	80.4	163	11	94.6	471	19	83.3	214	7	80.4	-	-	-
Nov	101	9	77.5	15	3	85.3	351	23	85.3	69	8	77.5	-	-	-
Dec	11	2	83.4	47	4	80.6	92	7	83.4	173	5	80.6	-	-	-

Conclusion

The present study reveals that there is a continuous need for screening sugarcane varieties for better quality, yield as well as disease resistance. The use of resistant varieties has been proved to be the best approach for the management of red rot disease in sugarcane. The different levels of resistance to red rot disease available in different sugarcane genotypes can be used in breeding programmes. The genotypes exhibiting both high brix % as well as red rot resistance can be released as new varieties or can be used as suitable parents in breeding programmes for evolving high yielding, high sugared and red rot resistant varieties.

References

[1] Viswanathan R and Padmanaban P (2008). Hand book on sugarcane diseases and their management. *Sugarcane Breeding Institute*, ICAR. 72p.

[2] Lewin HS, Natarajan and Rajan S D (1976). Control of sugarcane red rot (*Physalospora tucumanesis* Speg) by chemotherapy. *Sugarcane Pathol. Newslett.*, 17: 17-20.

[3] Agnihotri V P (1990). Diseases of sugarcane and sugarbeat, Oxford and IBH, Pub. Co. Pvt. Ltd., New Delhi- 483p.

[4] Vijai Singh, Joshi B B, Awasthi S K and Srivastava S N (2008). Eco-friendly management of red rot disease of sugarcane with *Trichoderma* strains. *Sugar Tech.* 10 (2): 158-161.

[5] Satyavir, Singh N, Virk K S, Nageswararao G V, Singh H and Misra S R (2001) Pathogenic variability in sugarcane red rot system. *In Proc. National Symp.- Role of resistance in intensive agriculture*, (Eds) S. Nagarajan and O.P. Singh, Kalyani pub., Ludhiana - 109–114 pp.

[6] Narendra Singh, Kumar S and Goraya SS (2000). Red rot disease scenario in Punjab state of India – an eye opener. *Indian Sugar*. 50 (8): 497 – 504.

[7] Chona B L (1954). Studies on the diseases of sugarcane in India. IV. Relative resistance of sugarcane varieties to red rot. *Indian J. Agric. Sci.*, 20: 363-385.

[8] Duttamajumder S K and Misra S C (2004). Towards an ideal method of inoculation for screening sugarcane genotypes against red rot caused by *Colletotrichum falcatum. Indian Phytopath*. 57: 24-29.

[9] Srinivasan K V and Bhat N R (1961). Red rot of Sugarcane criteria for grading resistance. *J. Indian Bot. Sci.*, 11: 566-577.

[10] Khan S H, Muhammad Shahid, Safurehman and Azher Mustafa (2009). Control of red rot disease of sugarcane through screening of varieties and seed dressing fungicides. *Pak. J. of Phytopath.* 21 (1): 61-65.

[11] Gupta A K and Vivek Yadav (2009). Evaluation of different sugarcane varieties for resistance against red rot disease. *Environment and Ecol.* 27 (3): 1006-1008.

[12] Malathi P and Viswanathan R (2012).Variation in *Colletotrichum falcatum* – Red rot pathogen of sugarcane in relation to host resistance. *Sugar Tech.* 14 (2): 181- 187.

[13] Rahman S (1996). Studies on the red rot of sugarcane in Bangladesh. Ph.D. Thesis, Rajshahi University, Bangladesh - 146 pp.

[14] Baksha R, Alam R, Kamal M M, Podder BP and Rahman AB M M (2003). Screening of Different Sugarcane species (*Saccharum Officinarum* and *Saccharum spontaneum*) for Red Rot Disease Resistance. *Plant Pathology Journal,* 2: 111-113.

In: Current Status of Sugarcane Research in India
Editors: A. K. Tiwari, M. Lal and A. K. Singh
ISBN: 978-1-63463-458-8

Chapter 9

UTILIZATION OF TISSUE CULTURE DERIVED VARIATION IN SUGARCANE IMPROVEMENT

V. P. Sobhakumari*
Tissue Culture Laboratory, Sugarcane Breeding institute,
Tamil Nadu, India

ABSTRACT

The present review highlight some of the developments in the field of *in vitro* culture induced variations that are evolving in the recent years as novel strategies for use in sugarcane improvement. Tissue culture, a routine technique covering a variety of branches has been widely involved in crop improvement especially by inducing genetic changes or somaclonal variation and can become a part of plant breeding provided they are heritable and genetically stable. Sugarcane is considered to be an ideal crop for such applications because of the existence of high ploidy, the capacity of plants to tolerate chromosomal aberrations and the capacity of the deviant cell to differentiate into plants. It is necessary to understand the basis of somaclonal variation so that it can be used in improvement programmes without disturbing the genetic constituent of a clone.

Keywords: Tissue culture, Somaclonal variation, Sugarcane, Callus culture

INTRODUCTION

The assembly of genetic variability is vital to any plant breeding enterprise. Conventionally, plant breeders recombine the desired genes from crop varieties and released species by sexual hybridization, and develop new cultivars with the desirable traits such as high yield, resistance to disease, insect and pests and drought. They are now faced with an even greater challenge to sustain food production for the ever growing human population. The adoption of new technologies such as plant tissue culture and transformation methods may

* Email for correspondence: vpsobhakumari@rediffmail.com.

help in achieving some of the goals to increase food production. There is a great potential of cell and tissue culture techniques in plant improvement, provided plants can be readily regenerated in large numbers.

Plant tissue culture methods offer a rich scope for the creation, conservation and utilization of genetic variability for the improvement of agricultural crops. Plants derived from tissue culture are termed somaclones and variations displayed by such somaclones are called somaclonal variations [1]. It can result in a range of genetically stable variation, useful in crop improvement, similar to that induced with chemical and physical mutagens. Somaclonal variation is unpredictable in nature, and can be both heritable (genetic) and non heritable (epigenetic). Somaclonal variation, cellular selection and early rapid screening of regenerants collectively provide a powerful option for plant improvement and this may be the best approach to plant improvement outside of conventional breeding. The mutational events can be triggered and the resulting genotypes preserved by clonal multiplication.

The recovery of somaclonal variation can be enhanced by: 1. callus and suspension cultures for several cycles, 2. regeneration of large number of plants from long term cultures, 3. screening of desirable plants and their progenies, 4. testing of selected somaclones in the subsequent generations for genetic stability, and 5. multiplication of genetically stable somaclones for developing new cultivars.

Basis of Somaclonal Variation

The occurrence of somaclonal variation is associated with gene mutations, chromosomal rearrangements and recombination, DNA methylation, altered sequence copy number, transposable elements etc and seems to be influenced by genotype, explant type, culture medium and age of the donor plants [2, 3, 4]. Apart from the plant type, the number of subcultures is another important aspect that can lead to more variation. Some reports revealed the effect of duration of callus culture on the accumulation of genetic alterations [5]. Tissue culture system itself act as a mutagenic system because cells experience traumatic experiences from isolation, and may reprogramme during plant regeneration which are different than under natural conditions. Reprogramming or restructuring of events can create a wide range of variation in newly regenerated plants [6,7].

Somaclonal Variation at the Chromosomal Level

Chromosomal variation has been observed in several tissue culture-derived plant species, and their progenies. The high ploidy and high-chromosome explants show more variability than low ploidy and low chromosome number species. Ploidy in tissue culture derived plants generally results from endopolyploidization or nuclear fusion [8, 9]. The altered karyotypes in somaclones include chromosomal rearrangements as well as aneuploidy and euploidy. Anueploidy may be caused by non disjunction, aberrant spindles, lagging chromosomes, chromosome breakage that produces dicentric and acentric chromosomes. Normal cell cycle controls, which prevent cell division before the completion of DNA replication, are presumed to be disrupted by tissue culture, resulting in chromosomal breakage [10]. Chromosome breakage and its consequences (deletions, duplications, inversions, and translocations) cause aberrations [11]. Chromosome breakage may also create mutations directly through 'position effect' or alteration in gene expression from chromosomal rearrangement. Furthermore, altered levels of DNA methylation can trigger chromosomal breakage. The age of callus also affects the frequency of chromosomal aberrations. In general, as the callus gets older the frequency of chromosomal instability increases.

Transposable Elements

Transposable element activation is another type of variation induced by tissue culture. Groose and Bingham [12] identified an unstable flower color mutation, which acted like a transposable element induced mutation. However, the implication of transposable elements is yet to be proved. Kaeppler et al. [13] suggested that transposable elements probably account for a relatively small proportion of tissue culture -induced variation.

Molecular Variation

Molecular variation in tissue culture derived plants has been characterized at DNA and protein level. Variation at the DNA level has been most extensively studied using restriction enzyme analysis. In most cases, changes in restriction pattern appeared as altered fragment size, rather than addition or loss of restriction fragment [13]. Studies using restriction enzymes sensitive to 5- methyl cytosine modification have shown extensive and frequent changes in RFLPs [14]. The most frequent variation at the protein level has been shown in grain storage proteins and isozymes. Characterized variation can be summarized into three categories: a) altered electrophoretic mobility, b) loss or gain of protein band, and c) altered levels of specific proteins.

DNA Methylation

DNA methylation has been associated with gene altered expression in numerous plant and animal species [13]. The direct role of DNA methylation in gene expression is still a subject of debate even though cytosine methylation is correlated with modified gene expression in plants and animals. Methylation can enhance quantitative trait variation because several genes can be affected simultaneously. Methylation of a gene inactivates its transcription and thereby controls gene expression during somatic embryogenesis. The resulting variation from DNA methylation is 'epigenetic'.

Potentiality of Somaclonal Variation

The potential usefulness of *in vitro* selection has increased dramatically with clear evidence of expression of selected traits in regenerated plants. Somaclonal variation has advantages as well as disadvantages. It is unpredictable in nature and can be heritable or non heritable. Somaclonal variation can become a part of plant breeding provided they are heritable and genetically stable. Since somaclonal variation can broaden the genetic variation in crop plants, many plant characters can be altered, including plant height, yield, number of flowers per plant, early flowering, grain quality in leguminous crops, resistance to diseases, insect and pests, cold, drought and salt tolerance. A wide variation in quantitative traits such as plant height, plant yield, tiller number and oil content was also reported through the induction of somaclonal variation.

Several reports have indicated the value of the selected somaclones in plant breeding. The question arises whether we can use this approach for improving agronomically important polygenic traits. The successful release of the high yielding and shattering resistant Indian mustard variety, developed through somaclonal variation suggests that it should be possible to develop new cultivars with improved polygenic traits.

Somaclonal Variation in Sugarcane

The potential usefulness of somaclonal variation for plant improvement first became apparent in sugarcane. The major area of utilization of tissue culture in sugarcane improvement is for the production of somaclones from callus cultures of commercially important varieties to rectify their specific defects. The highly polyploid nature of sugarcane coupled with chromosome numerical variation in different cells of same tissue is an added advantage in creating somaclonal variation for a wide spectrum of characters. Research on sugarcane tissue and cell culture was started in Hawaii in 1961 by Nickell [15]. Callus induction and subsequent shoot differentiation were first reported by Heinz and Mee [16]. Variation in sugarcane derived from cell cultures was reported in 1969 by researchers at Hawaiian Sugar Planters Association Experiment Station [17,18]. The initial site of callus cells in young leaf and sub apical meristem explants was first identified by Liu and Chen [19] who were the first among sugarcane cell culturist to attain several high yielding and high sucrose callus derived clones. Later several somaclones were developed through tissue culture with improved productivity and eliminating certain minor defects like spines, leaf drying and disease susceptibility. Somaclonal variation has been frequently reported in sugarcane [20, 18, 21, 22, 23] and has been intensively evaluated as a means for improving disease resistance [24, 25, 26, 27, 28, 29, 30], insect resistance [31], yield characteristics [32] and plant morphology [17, 33] of specific clones.

Induced variability may be either desirable, as in breeding, or undesirable as in germplasm conservation, depending on the research application. Among the *in vitro* methods commonly used to propagate sugarcane, shoot tip culture reportedly induces less variation than direct regeneration and callus culture. In this genetic stability is expected because the incipient shoot has differentiated *in vivo* and only elongation and root differentiation are required. Direct or rapid regeneration minimizes somaclonal variation by reducing time in culture and minimizing or eliminating callus formation. Sugarcane plants arising from the cell differentiation and redifferentiation process during callusing may be highly variable for chromosome number and agronomic characteristics [34, 1]. The frequency of abnormal regenerants can increase with callus age.

It is evident that callus culture can be utilized as an adjunct to breeding programmes to rectify specific defects and to induce novel variability without disturbing the fine genetic balance of the genotype. This approach was utilized to rectify the specific defects of the widely adapted sugarcane varieties as well as the genotype at the final stage of selection which was rejected due to specific defects. Somaclonal variation noticed in the callus derived plants of sugarcane cultivars with respect to smut disease resistance was studied by Sreenivasan and Jalja [35]. The study clears that chromosome numerical variation is not a pre requisite to somaclonal variation but it only enhances frequency of extreme variance which could be effectively utilized to improve smut disease resistance in sugarcane cultivar, without adversely affecting other desirable traits. The comparative study with respect susceptibility to rust and yield characteristics of sugarcane callus culture has showed that the effect of rust resistance persisted *in vitro* derived plants even after two vegetative propagations [36].

Sugarcane somaclones regenerated through callus of variety CoS 91279 showed wide variations for red rot resistance against four isolates of *Colletotricum falcatum*. Out of 42 somaclones tested only three were found moderately resistant by plug method of inoculation. Rest of the somaclones showed varying degree of susceptibility [37]. The effect of phytotoxin of *Colletotricum falcatum* on sugarcane tissue culture was studied by Mohanraj et al. [38] who revealed the possibility of using the pathogen toxin to produce red rot resistant genotypes of sugarcane. Salt tolerant somaclones of sugarcane performed better in characteristics like number of tiller/plant, stem height and number of nodes/stem. The somaclones performed less in characteristics of growth of stem and brix percentage [39].

At Sugarcane Breeding Institute, Coimbatore, tissue culture methodology has been well developed (Figure 1) and several somaclones were developed through this method with improved productivity and eliminating certain defects like leaf spines (Co 7717), leaf drying (Co 7704), disease susceptibility (CoA 7601- rust, Co C 671 – smut), Salinity (Co 8021 and Co 62175) etc.

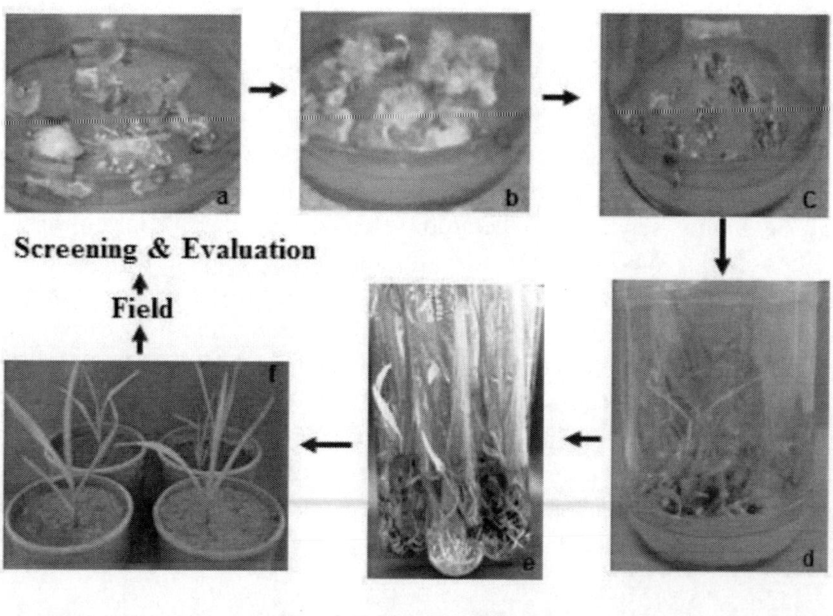

Figure 1. Steps involved in sugarcane tissue culture a) Leaf bit explants; b) Callus induction from explants; c) Shoot regeneration from callus; d) Shoot multiplication e) *In vitro* rooting f) Hardening.

The results of genetic variation reported by various researchers may be a reflection of different genotypes and experimental conditions employed. At our Institute, *In vitro* culture study has been conducted with different clones of intergeneric and interspecific hybrids of sugarcane and identified the genetic and nongenetic factors that influence the callus induction and regeneration [40]. Improved method of *in vitro* culture in sugarcane hybrids has also been standardized with partial desiccation of calli [41]. Induction of flowering on a *Sorghum* x *Saccharum* hybrid could be achieved through gamma irradiation of calli [42]. In this intergeneric hybrid, ploidy could be increased by *in vitro* colchiploid production and it has been confirmed through cytological analysis (Figure 2) [43].

Figure 2. *In vitro* induction of colchiploids in *Sorghum* x *Saccharum* hybrid and cytological analysis of control and colchiploids a) Original *Sorghum* x *Saccharum* hybrid (SSH 1); b) Somatic chromosome number of the hybrid (2n=66); c and d) Calli induction and shoot differentiation from calli; e) Stomata of control f) Stomata of colchiploid; g) Colchiploid plants; h-j) Somatic chromosome numbers of different colchiploids.

Amphiploids could be obtained in another intergeneric hybrid of *Erianthus* x *S. spontaneum* through colchicine treatment of calli (unpublished report). Since *in vitro* tissue cultures may be highly sensitive to heterotic effects they may help in early and rapid identification of superior hybrid combinations for agronomic characters. Since the vigor at the cellular level may be related to vigor for agronomic characteristic the technique can also be used as a tool for early screening for the choice of best parents and hybrids in plant breeding programmes [44]. Some of the Co canes developed at Sugarcane Breeding Institute viz., Co 92007, Co 92029, Co 93005, Co 94003, Co 94012, Co 99011, Co 98016 and Co 99012 are somaclones. Co 94012 has been released for cultivation in Maharashtra which is found to give high sugar recovery. It is quite clear that somaclonal variation can be induced easily in sugarcane to develop the somaclones with improved traits for the improvement of varieties particularly those with single defects.

FUTURE PROSPECTS

Somaclonal variation has a vast potential for inducing genetic variation in a crop, but there is a need to emphasize on its use in the crop improvement. To select a somaclone with desirable trait, it is essential to produce large population of plants. Somaclonal variation results in the production of new genotypes with a little or no change in the original genome. Molecular markers such as RAPD, RFLP, AFLP and microsatellites are appropriate tools to identify genetic and epigenetic somaclones.

It seems feasible to understand the molecular basis of somaclonal variation so that it can be used without any loss of genetic trait. Plant breeders need to be convinced that the stable somaclones are safer to use in breeding new varieties. A reliable molecular technique needs to be developed in order to identify genetic variation at an early stage of plant development. Different molecular and biochemical techniques can be employed to detect the full spectrum of somaclonal variation that may arise by mechanisms that range from chromosome rearrangements or breakage and activation of transposable elements to point mutations.

Extensive application of tissue culture in sugarcane improvement will await the standardization of appropriate screening methods at cellular level or at least at the early stage of plant differentiation. *In vitro* selection will save the time taken for selecting the clones with disease resistance and tolerance to abiotic stresses through conventional methods. *In vitro* selected putative variants should be finally field tested to confirm the genetic stability of the selected trait.

Gene transfer from related genera like *Erianthus, Sclerostachya* etc. to sugarcane is assuming importance to evolve varieties suitable for cultivation under marginal land with low input and for fiber and biomass. Because of the prevalence of autosyndetic chromosome pairing interspecific and intergeneric hybrids of *Saccharum*, chromosome segmental exchange between species or genera never takes place normally. Tissue culture can be an ideal system for inducing chromosome interchanges. The system can also be used as an efficient method to produce tetraploids to restore fertility in sterile hybrids.

Though some attempts have been made for the induction of haploid through anther culture, true haploid production has not yet achieved. If polyhaploids could be produced in large numbers, it may be of great use in genetical and cytogenetical studies.

In recent years, distant hybridization has been a fascination of sugarcane scientists. Hybrids were successfully produced with distant genera like sorghum and corn, however, an attempt made with bamboo was unsuccessful. Fertilization of sugarcane and bamboo gametes and abortion of the embryo at an early stage have been observed. Embryo culture and protoplast culture may prove to be useful in obtaining such distant hybrids in sugarcane.

CONCLUSION

The major area of utilization of tissue culture in sugarcane improvement is for the production of somaclones from callus culture of commercially important varieties to rectify their specific defects. Critical to develop tissue culture derived variants in sugarcane is the predictability and stability of variations. Understanding and implementing the factors affecting these variations can possibly overcome this problem. Sugarcane is a suitable candidate for the application of tissue culture induced variation because of its polyploid nature and high regeneration capacity. Thus this system can be applied in sugarcane breeding programmes as a complimentary system for the development of improved subclones for commercial purposes, parental lines, genetic stocks and energy cane.

REFERENCES

[1] Larkin P J and Scowcroft S C (1981). Somaclonal variation-a novel source of variability from cell culture for plant improvement. *Theor. Appl. Genet.* 60: 197-214.

[2] Veilleux R E and Johnson A A T (1998). Somaclonal variation: Molecular analysis, transformation, interaction and utilization. *Plant Breed Rev.* 16: 229-268.

[3] Jain S M, Brar DS and Ahloowalia BS (Eds) (1998). Somaclonal variation and induced mutations in crop improvement . Kluwer Academic Publishers, UK.

[4] Winkelmann T and Serek M (2005). Genotypic differences in callus formation and regeneration of somatic embryos in *Cyclamen persicum* Mill. *Euphytica.* 144: 109-177.

[5] Kuznetsova O I, Ash OA and Gostimsky SA (2006). The effect of duration of callus cultureon the accumulation of genetic alterations in pea, *Pisum sativum* L. *Russian J. Genet.* 42: 555-562.

[6] Jain S M (2000). Mechanisms of spontaneous and induced mutations in plants. *Radiation Res Vol. 2. Cong. Proc.* Pp 255-258.

[7] Jain S M (2001). Tissue culture derived variation in crop improvement. *Euphytica.* 118: 153-166.

[8] Sunderland N (1977). Nuclear cytology. In: H.E. Street (Ed.) *Plant tissue and cell culture.* Pp177-205. Blackwell. Oxford.

[9] Bayliss M W (1980). Chromosomal variation in tissue culture. *Intern Rev. Cytol. Supple* IIA 113- 144.

[10] Plillips R L, Kaeppler SM and Olhoft P (1994). Genetic instability of plant tissue cultures: break down of normal controls. *Proc. Natl. Acad. Sci. USA* 91: 5222-5226.

[11] Duncan R R (1997). Tissue culture induced variation and crop improvement. *Adv. Agron.* 58: 201-240.

[12] Groose R W and Bingham ET (1986). An unstable anthocyanin mutation recovered from tissue culture of alfalfa. 1. High frequency of reversion upon reculture. 2. Stable non revertants derived from reculture. *Plant cell* rep 5: 104-110.

[13] Kaeppler SM, Phillips RL and Olhoft P (1998). Molecular basis of heritable tissue culture induced variation in plants. In: S M Jain D S Brar and B S Ahloowalia (Eds.) Somaclonal variation and induced mutations in Crop improvement, pp 467-486. Kluwer Academic Publishers, Dordrecht.

[14] Kaeppler S M and Phillips RL (1993). DNA methyletiopn and tissue culture induced variation in plants. *In Vitro Cell dev. Biol.* 29: 125-130.

[15] Nickell L G (1964). Tissue and cell culture of sugarcane, Another research tool. *Hawaii Pl. Rec.* 57: 223-229.

[16] Heinz D J and Mee G W P (1968). Tissue callus differentiation and regeneration of plants in Saccharum spp. *Agron Abstr.* pp 10.

[17] Heinz D J and Mee GWP (1969). Plant differentiation from callus tissues of *Saccharum* species. *Crop. Sci.* 9: 346-348.

[18] Heinz D J and Mee GWP (1971). Morphologic, cytogenetic and enzymatic variation in *Saccharum* species hybrid clones derived from callus culture. *Amer. J. Bot.* 58: 257-262.

[19] Liu M C and Chen WH (1982). Application of cell tissue culture techniques for sugarcane improvement. *Ann. Rep. Res. Dvpt Council* 14-15, Taiwan Sugar Corp, Taiwan.

[20] Kresovich S, McGee RE, Draweand HJ and Rivera JL (1986). Variability of agronomic characters in populations of tissue culture derived and vegetatively propagated sugarcane. *Proc Int. Soc. Sugarcane Cane technol.* 19: 528-532.

[21] Liu M C and Chen WH (1976). Tissue and cell culture as aids to sugarcane breeding. I. Creation of genetic variation through callus culture. *Euphytica* 25: 393-403.

[22] Lourens A G and Martin FA (1987). Evaluation of in vitro propagated sugarcane hybrids for somaclonal variation. *Crop. Sci.* 27: 793-796.

[23] Nagai C, Ahloowalia BS and Tew TL (1991). Somaclonal variants from an intergeneric hybrid: *Saccharum spp* hybrid x Erianthus arundinaceum. *Euphytica* 53: 193-199.

[24] Heinz D J, Krishnamurthi M, Nickell L G, and A Maretzki (1977). Cell, tissue and organ culture in sugarcane Improvement. In: Reinert J, Bajaj YPS (Eds), Applied and fundamental aspects of plant cell, tissue and organ culture. Spinger-Verlag, Berlin Heidelberg, New York. 3-248.

[25] Krishnamurthi M and Tlaskal J (1974). Fiji disease resistant Saccharum officinarum Var.pindar subclones from tissue cultures. *Proc. Int. Soc. Sugarcane Technol.*15: 130-137.

[26] Bonnel E, Peros JP and Girard JC (1988). Evaluation de la resistance a la gommoseet du rendement agronomique de somaclones de canne a sucre issus du cultivar R 472. *Agron. Trop.* 43: 252-255.

[27] Larkin P J and Scowcroft SC (1983). Somaclonal variation and eyespot toxin toletance in sugarcane. *Plant Cell Tiss. Org. Cult* 2: 111-122.

[28] Sreenivasan J, Sreenivasan TV and Alexander KC (1987). Somaclonal variation for rust resistance in sugarcane. *Indian J. Genet Plant Breed.* 47: 109-114.

[29] Ramos Leal M A, Maribona R H, Ruiz A, Korvena S, Canales E, Dinkova T D, Izquierdo F, Coto O and D Rizo (1996). Somaclonal variation as a source of resistance to eyespot disease of sugarcane. *Plant Breeding.* 115: 37-42.

[30] Sobhakumari V P and Malathi P (2013). *In vitro* induction and evaluation for smut resistance in Sugarcane. Proceedings of International Conference on Advances in Biotechnology and Patenting. Bharathidasan University, Trichy (18-21, Feb, 2013). Page 24.

[31] White W H and Irvine JE (1987). Evaluation of variation in resistance to sugarcane borer (Lepidoptera: Pyralidae) in a population of sugarcane derived from tissue culture. *J. Econ. Entomol.* 80: 182-184.

[32] Liu M C and Chen WH (1978). Tissue and cell culture as aids to sugarcane breeding. II Performants and yield potential of callus derived lines. *Euphytica.* 27: 273-282.

[33] Nagai C, Ahloowalia B S, Heinz D J and Tew TL (1986). Colchicine-induced anueploids from cell culture of sugarcane. *Euphytica* 35: 1029-1038.

[34] Evans D A, W R Sharp and Medina-Filho HP (1984). Somaclonal and gametoclonal variation. *Am. J. Bot.*71: 759-774.

[35] Sreenivasan T V and Jalaja N C (1982). Production of subclones from the callus culture of *Saccharum-Zea* hybrid. *Plant Sci. Lett.* 24: 255-259.

[36] Perose J P, Bonnel E, Roques D and Paulet F (1994). Effect of *in vitro* culture on rest resistance and yield in sugarcane. *Field Crops Res* 37: 113-119.

[37] Singh A, Lal M, Singh MP, Lal K and Singh SB (2000). Variations for Red rot resistance in somaclones of sugarcane. *Sugar Tech.* 2 (1 and 2): 56-58.

[38] Mohanraj D, Padmanaban P and Karunakaran M (2003). Effect of phytotoxin of Colletotrichum falcatum Went. (Physalospora tucumanensis) on sugarcane tissue culture. *Acta phytophthologica Entomologica Hungarica*, 38: 21-28.

[39] Khan S T, Khan MA, Ahmad HK, Khan RD and Zafar Y (2004). Somaclonal variations in sugarcane through tissue culture and subsequent screening for salt tolerance. *Asian J. Plant Sci.* 3 (3): 330-334.

[40] Sobhakumari V P, Sreedivya MJ and Sanu Mary Abraham (2010). Genetic and non-genetic factors affecting callus induction and regeneration in sugarcane. *Gregor Mendel Foundation Journal* 1 (1and2): 5-9.

[41] Sobhakumari V P, Janani AVR and Nithiyanantham R (2011). Improved method of *in vitro* culture in hybrids of sugarcane. Proc. National seminar on Genetics, Breeding and Biotechnology. *Calicut.* p.21.

[42] Sobhakumari V P and Nair NV (2005). Induction of flowering in a *Sorghum* x *Saccharum* hybrid through gamma irradiation of calli. *Cytologia.* 70 (4): 393- 397.

[43] Sobhakumari V P and Nair NV (2007). *In vitro* induction of colchiploidy in an intergeneric hybrid of *Sorghum x Saccharum. Cytologia*, 72 (3): 303-308.

[44] Sobhakumari V P and Mathew SD (2009). Effect of Hybrid vigor on callus induction and regeneration of Sugarcane. *Cytologia* 74(1)71-77.

In: Current Status of Sugarcane Research in India
Editors: A. K. Tiwari, M. Lal and A. K. Singh

ISBN: 978-1-63463-458-8

Chapter 10

SSI (Sustainable Sugarcane Initiative) Technology: A Way Forward for Enhanced Cane Production and Economic Returns

M. Mohanty**, *P. K. Nayak and S. S. Nanda
Sugarcane Research Station (Odisha University of Agriculture and Technology), Nayagarh (Odisha), India

Abstract

Sugarcane, being an input and labour intensive crop is at high risk due to high input costs, labour unavailability, low irrigation potential, aberrant climatic situations and incidence of pest and diseases. About 10% of sugarcane produced is used by the growers as seed material for planting in subsequent year. Despite high seed rate, close planting can only support a population of 62,000 canes/ha due to high mortality while competing for sunlight and nutrients with lesser number of tillers and few millable canes. To address these problems, a package of simple agricultural innovations called Sustainable Sugarcane Initiative (SSI) is applied for sugarcane farming using less inputs, seed, water and fertilizers. Use of bud chips instead of 3-bud setts as planting material and transplanting seedlings raised from bud chips with wider row spacing is the basic protocol of SSI. Single-bud chips, carefully removed from healthy canes are used for raising the nursery. Only 50-75 kg of bud chips are used for a hectare of crop and the remaining canes could be sent for crushing. It is important to treat the bud chips with various organic or chemical solutions before planting to avoid infestation. The buds are placed in the cones of plastic trays along with the coco-pith (coconut coir waste) and well powdered FYM/ vermicompost mixture (3:1 ratio). Through this method, a high percentage of germination can be achieved within a week, based on the agro-climatic conditions. About 1400 canes are needed to get 14,000 buds sufficient for 1 hectare plantation of seedlings with 4 X 2 ft spacing. About 20-25 day-old seedlings can be removed and planted in the main field. Application of organic manures like FYM/compost/well decomposed press mud and use of bio fertilizers like *Trichoderma*, PSB, *Azotobacter* and *Pseudomonas* are encouraged. In the conventional method 40,000

* Corresponding author: Email; mmohanty877@gmail.com.

three-bud setts are planted to achieve a normal population of 1,00,000 canes/ha. With the SSI method of sugarcane cultivation, wide spacing of 4-5 X 2 ft in the main field gives 1,12,000 to 1,37,000 millable canes because of more tillering. This wider spacing in SSI cultivation reduces the seed usage to mere 12,000 to 13,750 bud chips grown seedlings, compared to 1,20,000 buds in three budded setts, 80,000 buds in the two-bud setts and 40,000 buds in one-bud setts in conventional cultivation. The wider spacing between the rows provides ample scope for intercropping within standing crop of sugarcane. Crops like green gram, cowpea, gram, potato, onion, wheat, coriander, lady's finger and melons can be effectively taken up as intercrop with sugarcane. An on-farm trial was conducted for two consecutive cropping seasons of 2011-12 and 2012-13 using a participatory approach on cultivators' fields in Odisha revealed that the seedling survival rate was 88 % in SSI technology as compared to only 55.81 % bud germination in conventional method. Average number of millable canes was higher in SSI technology as compared to conventionally grown crop. Higher plant stand along with higher yield attributing characters resulted in higher cane yield of 105.0 t/ha in SSI technology as compared to 89.0 t/ha under traditional three bud setts planting.

Keywords: Conventional method, Sustainable Sugarcane Initiative, seedling transplanting, wider spacing, low cost cultivation, water saving technology, higher cane yield, intercropping, higher net return

INTRODUCTION

Sugarcane (*Saccharum* sp. complex) is one of the most efficient converters of solar energy into sugars and other renewable forms of energy. In India, It is cultivated in an area of 5.025 m ha with a total production of 342.56 million t of sugarcane and 26.5 million t of sugar at an average productivity of 68.1 t/ha [1]. About 50 million people depend on this crop, including the employment generated by around 570 sugar factories and other related industries. It not only produces sugar but also every by product has economic uses like fodder, paper and production of bio-fuels. In a typical sugar mill, 100 tonnes of sugarcane on an average produces 10 tonnes of sugar, 4 tonnes of molasses from which ethanol is produced, 3 tonnes of press mud which is converted into organic manure, 30 tonnes of bagasse used to yield 1,500 kw electricity besides 30 tonnes of cane tops and leaves generally left in the field. Sugarcane cultivation in India is in crisis due to stagnant productivity (65–70 tonnes/ha) level during last decade. India contributes about 12 percent of world sugar production with a total investment of $11 billion, which is no longer limited to sugar but also includes the co-generated power and ethanol sector as well.

Being an input and labour intensive crop; high input costs, labour unavailability at peak demand season, low irrigation potential, aberrant climatic situations and incidence of pest and diseases are the reasons for low productivity of sugarcane. India, with the second largest area under sugarcane cultivation in the world, around 5.025 million ha, is in big trouble. In countries like India, it is the small landholding farmers who cultivate crops predominantly and the costs of cane cultivation have risen alarmingly for seed/planting material, manures and fertilizers, irrigation, cultural practices and harvesting. In normal course, for commercial cultivation, a huge quantity (8-10 t/ha) of cane stalk cuttings having 3-bud pieces (25-30 cm long segments) are required for planting one hectare land. Such planting material ranges from 22 to 25% of the total production cost, and that is one of the major items of expenditure in sugarcane cultivation [2]. On an average, 10% of sugarcane produce is used as planting material. In other words, we are losing 1.5 million tonnes of sugar annually by burying it in the soil as a planting material [3].

Despite high seed rate, close planting can only support a population of 62,000 canes per hectare due to high mortality while competing for sunlight and nutrients with lesser number of tillers (6-8 / plant) and few millable canes (2-3 / clump). The average weight of canes is 0.75 kilogram, which under good circumstances ends up yielding about 60 to 80 tonnes/ha. Depleting water tables and absence of proper water management practices, shortage of labour as the traditional practices have high labour requirement poses great difficulty for sugarcane growers of our country.

To address these problems, a package of simple agricultural innovations called Sustainable Sugarcane Initiative (SSI) is applied for sugarcane farming using less inputs, seed, water and fertilizers inspired by SRI- System of Rice Intensification experience in rice cultivation. This is called 'More with Less' approach in agriculture. SSI has helped to improve the water productivity by 40 percent, the profits by 30 percent while reducing the ecological impact. SSI leads to healthier soil and plants supported by greater root growth and the nurturing of soil microbial abundance. In addition, it has been found in farmers' experience that using drip irrigation leads to a great saving of water, by as much as 80 percent. As such, SSI is becoming a focus for the industry, governments, as well as financial institutions for its scaling up.

Constraints Confronted by Sugarcane Farmers of India

- High cost of sugarcane cultivation and low productivity are serious threats the cane growers of India are grappling with.
- Higher seed rate due to closer row spacing.
- High rate of chemical fertilizers resulting in imbalanced nutrient management.
- High labour requirement in various cultural operations.
- Higher cost of irrigation due to escalating cost of electricity charges.
- The flooding method of irrigation is wasteful, causing huge strain on local ground water resources.
- Un-availability of labourers at peak demand season.
- Non-availability of situation specific cane varieties.
- Degeneration of new and promising sugarcane varieties after few years of their release.
- Depleting water tables, development of salinity or alkalinity in irrigated tracts, water stagnation during grand growth phase of the crop.
- Unpredictable climatic aberrations.
- Improper cultivation practices, negligence in plant protection measures, and other practices like mono-cropping, generally result in low productivity and ultimately translates into lower net return

On one hand, there is a lot of scope for the cane growers in view of growing demand for sugar and other by-products of sugarcane who are at risk due to decline in production and productivity due to various reasons. The average productivity of sugarcane is low with certain regions reporting yields as low as 40 t/ha only. Not only is the cane yield low; the sugar yield is typically less than 10 percent of cane weight, which is less than satisfactory given that yields of 14 percent of cane weight at the time of cutting (and sometimes even higher) are possible. There are large losses between cutting in the field and processing in mills.

The recent successes of SRI and SSI are a clear indication that the modern problems of water crisis, soil degradation, stagnant yields, high input costs in agriculture, loss of varieties, etc., can to some extent be addressed with some modifications to our existing knowledge systems. In the coming years, sugarcane farmers can reduce the seed material, by planting sugarcane in wider spacing which will facilitate intercropping and use of less water by adopting SSI technology of cane cultivation. SSI is not a new package of practices but a new way of thinking as well as cultivating that involves use of less seed cane, less water and optimum utilization of fertilizers and land to achieve more yield and profit for farmers and millers alike. It is an alternative way to the conventional seed cane, water and space intensive sugarcane cultivation. The challenges ahead call for capacity-building of farmers, service providers and research organizations on SSI.

Use of bud chips instead of 3-bud setts as planting material and transplanting seedlings raised from bud chips with wider row spacing has the following objectives envisaged at different points of time:

- Growing the crop from buds leaving the entire cane for commercial use thus saving large amount of cane from being buried.
- Treatment of seed materials become easier and more effective which in turn reduces the disease pest incidence.
- Transplanting of healthy seedlings in the main field thereby ensuring requisite plant stand.
- Nursery period of just one month, allowing a breathing spell for main field preparation.
- Use of bud chips for effective utilisation of precious seed cane in germplasm material.
- Easy transport of selections and test varieties across the country in varietal development programmes.

Over all, it is a holistic approach of 'more with less' with bud chip seedlings planted at wider row spacing which ultimately results in 'Sustainable Sugarcane Initiative - SSI', a better way of growing sugarcane with comparatively lower cost of cultivation.

The Principles of SSI

- Raising a nursery using single-bud chips from canes thus leaving the entire length of cane for commercial use.
- Transplanting young seedlings (25-30 days old).
- Maintaining wide spacing (4 to 9 X 2 ft) in the main field, thus gives scope for mechanization in sugarcane cultivation.

- Encouraging organic methods of plant nutrition, plant protection and other intercultural practices.
- Cultivation under sufficient moisture condition instead of flooding of fields.
- Scope for intercropping under wider row spacing for effective utilization of land and maintaining ground cover with enhanced economic returns.

Technology of Raising Sugarcane Crop through SSI Technology

Raising Nursery Using Single-Bud Chips

The Bud Chipper comprises a handle and a cutting blade fixed on a wooden/iron plank. Single-bud chips, carefully removed from healthy canes that are 7 to 9 months old and free from disease or pest infestations, are used for raising the nursery. Care should be taken while removing the dry leaves from the cane. It is preferable to remove the leaves manually to avoid damage of the buds. Hold the cane on the plank and adjust it in such a way that the bud is placed exactly below the cutting blade. When the handle is pressed, a single bud chip comes off from the cane and the entire length of cane is left for commercial use (Figure 1, 2, 3). About 400-500 buds can be chipped off in this way by two labourers in an hour. Take care to select only healthy buds while chipping and treatment. Only 50-75 kg of bud chips are used for a hectare of crop and the remaining canes could be sent for crushing. It is important to treat the bud chips with various organic or chemical solutions before planting to avoid infestation. Approximately 100 bud chips weigh 1 kg. Bud chips can be filled in a 5 kg bag for treatment. The bud treatment can be done in the following manner:

Chemical treatment	Organic treatment
Chlorpyriphos 2 ml + Bavistin 1.5 g + urea 10 g + lime 5 g / l of water	*Trichoderma* or *Pseudomonas* – 1 kg + Cow urine – 3 to 4 litres

Take a tub or drum (50 litres capacity), preferably made of aluminium or plastic. Pour 20 litres of water in the tub and dissolve the chemical or organic components as recommended above. Put the bud chips in a porous plastic/gunny bag or bamboo basket and immerse the bag/basket in the prepared solution for 20 minutes. Then the treated buds are shade dried. To raise the seedlings, the selected buds are placed individually in the cones of plastic trays along with the coco-pith (coconut coir waste) and well powdered FYM/ vermicompost mixture (3:1 ratio). Through this method, a high percentage of germination can be achieved within a week, based on the agro-climatic conditions. About 1400 canes are needed to get 14,000 buds sufficient for 1 hectare plantation of seedlings with 4 X 2 ft spacing even after deducting the wastage due to mortality in nursery and main field. Approximately 10 buds can be removed from each cane. Fill half of each cone in the tray with coco-pith and FYM/ vermicompost mixture. Place the buds in a slightly slanting position in half-filled cavities of trays (Figure 4). Do not press or push them hard. Ensure that the bud side faces up. Then cover the bud chips in the trays completely with coco-pith (Figure 5).

The soil of the nursery area should be drenched with Chlorpyriphos 50 EC (5ml/l) to control termites and care should be taken to avoid any weed growth. The nursery can also be set on roof tops or *verandah*. Bud treatment helps in 90 percent germination and subsequent health. For a 1 hectare plot using 4 X 2 ft spacing, 275 trays (each with 50 cones, to accommodate 13,750 pre-sprouted buds) and 375 kg coco-pith along with 125 kg vermicompost or FYM are sufficient to raise the seedlings needed (considering the mortalities in nursery).

Figure 1. Extraction of bud chips.

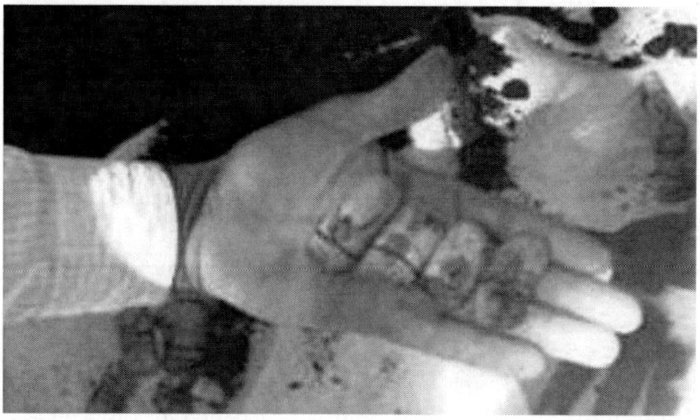

Figure 2. Healthy bud chips.

Figure 3. Canes left for crushing.

Figure 4. Planting of treated bud chips in the tray.

Figure 5. Covering of bud chips with coir pith + vermicompost.

Table 1. Number of seedlings required/ hectare at various spacing (including 15 % mortality)

Spacing	Number of seedlings/ha
4'X2'	13,615
5'X2'	10,890
6'X2'	9075
7'X2'	7777

Stacking of Trays

After covering, water all the trays lightly using rose can and then place them one above the other and finally, place an empty tray upside down on the top of the stack. This way, about 100 trays arranged in 4 sets (each set consisting of 25 trays) are to be placed together and wrapped tightly with black polythene sheets (Figure 6). Place small weights on the bundles and keep them closed for 5 to 8 days in the same position to create high temperature and good humidity.

Care should be taken to avoid water, air or sunlight entering into the trays by tightly covering them with polythene sheets. Keep a watch to prevent weed growth around the stacks.

Under warm temperature and high humidity generated inside the stacks, white root primordia will come out within 3-5 days and shoots will also appear in the next 2 to 3 days. After the buds are sprouted all the trays are to be removed from the polythene sheet on or between 5th and 8th day (based on the sprouting under climatic conditions) and are then kept side by side on the polythene sheets spread on the ground to facilitate watering and other nursery management practices. Based on the moisture content of the coco-pith, watering the trays has to be continued in the evenings for the next 15 days using rose cans. Shoots will start growing strong and leaves will start sprouting (Figure 7, 8). After the appearance of two leaves, application of water can be increased gradually depending on the moisture level in the trays.

Grading

During the 3-4 leaf stage (about 20-25 day-old seedlings), grading of the plants has to be done. Stop watering before a day of transplanting to loosen the coco-pith in the trays as this will enable easy removal of the young seedlings from the trays. Plants of similar height and vigour can be removed and placed in one tray. This way, healthy plants (Figure 9) are selected and damaged or dead plants can be eliminated which ensures the desired plant population in the main field.

Main Field Preparation

Main field preparation for sugarcane starts with clearing the preceding crop residues. Tillage operations can be carried out using harrows or rotavator.

The operations are to be repeated to make the soil bed free from clods, weeds and crop residues. After the tillage operation, the field should be deep ploughed using a tractor.

Application of Organic Manure

The SSI method encourages application of organic manure as much as possible as this enhances the macro and micro nutrient availability in the soil in an eco-friendly way, besides enhancing the use efficiency of chemical fertilizers side by side protecting the soil and environment from degradation and other hazardous effects.

Organic manure like FYM/compost/well decomposed press mud (@ 20-25 tonnes/hectare) is to be incorporated in the main field before the last ploughing. Bio fertilizers like *Trichoderma*, PSB, *Azotobacter* and *Pseudomonas* (10-12 kg in 500 kg of FYM/ha) can be mixed with the organic manures and applied in furrows before planting. This will control soil pathogens and improve the soil fertility to realize higher yields.

Figure 6. Wrapping up of trays after planting.

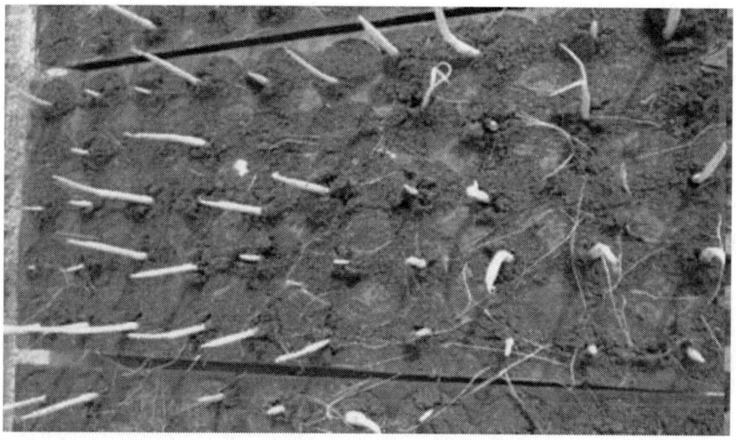

Figure 7. Sprouting of buds.

Figure 8. Development of seedlings.

Figure 9. A healthy seedling.

Transplanting of the Seedlings in the Main Field

The ideal age for transplanting young seedlings from the nursery to the main field is 25 to 30 days, as they will establish and grow better, with minimum loss due to transplantation shock (Figure 10, 11). The zigzag method of planting can be followed to utilize more space and achieve maximum tillers. For better access to sunlight, follow a north–south direction of planting. However, the slope of the field should also be taken into consideration. Seedlings are to be planted in the moistened soil in the furrows with a gentle thrust (Figure 12). Do not place the seedlings in the middle of furrows; this will hinder the root growth. To moisten the soil, irrigate the field one or two days before transplanting. Similarly, irrigation is to be done immediately after planting.

In the conventional method 40,000 three-bud setts (1, 20,000 buds) are directly planted in the field to achieve a normal population of 1, 00,000 canes per hectare. With the SSI method of sugarcane cultivation, wide spacing of 4-5 X 2 ft in the main field gives 1,12,000 to 1,37,000 millable canes because of more tillering. This wider spacing in SSI cultivation reduces the seed usage to mere 12,000 to 13,750 bud chips grown seedlings, compared to 1,20,000 buds in three budded setts, 80,000 buds in the two-bud setts and 40,000 buds in one-bud setts in conventional cultivation. A plant-to-plant distance is maintained at 2 ft cm within rows. Wider spacing helps in easy penetration of sunlight and air which helps in healthy growth of seedlings and controls pests and pathogens to some extent. Maintaining at least 4 - 5 ft distance between rows facilitates mechanical operations in the fields.

Water Management

Flooded condition during the crop formation stage will actually hinder the growth of the plant. It is always better to provide plants with sufficient quantity of water on time rather than continuously flooding the field. In the conventional flooding method, more water is always applied than the crop's biological demand which affects the crop's growth. Irrigation is normally applied once in 10 days during the tillering period (36-100 days), once in 7 days during the grand growth period (101-270 days) and once in 15 days during the maturity period (from 271 days till harvest).

Figure 10. Nursery raised from bud chips.

Figure 11. Seedlings taken for transplanting.

Figure 12. Seedling transplanting in pre-irrigated field.

Furrow irrigation helps in proper application and saving of water. Alternate furrow irrigation means irrigating the furrows with odd numbers initially, followed by irrigating the furrows with even numbers after 7 to 15 days, as per the moisture content of the soil and the age of the crop. This will ensure saving of water up to 50 percent. Drip irrigation can be practiced more effectively in SSI due to wider spacing and the planting of single seedlings. Water requirement for sugarcane is usually an average of 150 lakh litres/ha for a full season including rainfall. However, in the conventional method of flood irrigation, 200 lakh litres/ ha of water is applied by irrigation alone. In the drip system, irrigation efficiency improves by up to 90 percent and water is saved up to 40-70 percent. Consumption of electricity is also reduced. Furrow and alternate furrow irrigation can be followed to save water up to 50 percent. With SSI, about 5 irrigations can be saved as the germination period (up to 35 days) is spent in the nursery.

Fertilizer Application

Soil testing is a pre-requisite to know the nutrient status and for enriching the soil accordingly. If there is no such facility, then NPK can be applied at the rate of 208 kg N, 60 kg P and 120 kg K per acre, respectively, through inorganic or organic methods. Inorganic fertilizers like Urea, Di-Ammonium Phosphate (DAP), Muriate of Potash (MoP) and Ammonium Sulphate can be applied to achieve the above-mentioned nutrient requirement where supplies of organic nutrients and material are insufficient.

Table 2. Fertilizer dosage recommended for field application (per hectare)

Dosage	Days after planting	Fertilizers and their quantity (kg)			
		DAP (Diammonium phosphate)	MoP (muriate of potash)	Urea	Ammonium Sulphate
Basal	0	87	50	-	-
1st top dressing	30	50	-	62	-
2nd top dressing	60	-	75	188	-
3rd top dressing	90	-	75	250	-
Final top dressing	300	-	-	-	125
Total	480	137	200	500	125

The most appropriate method of applying fertilizers is by mixing them with organic manures, *neem* cake *etc.* and spot applying them through furrows at the root zone (2-3 inches away from roots). This will enable gradual release of nutrients supported by microbial activities. Applied fertilizers should be covered immediately with soil to avoid losses like volatilization. It is better to irrigate the furrows once the applied fertilizer is covered well with the soil. It is generally not good to apply fertilizers beyond 120 days, as this might reduce the cane quality. It is best to apply the fertilizers through drip irrigation (fertigation), which increases the fertilizer use efficiency of the crop and saves much of the input cost to the farmer. The recommended quantity of fertilizers can be applied in split doses (basal, 30, 60, 90 and 120 days after planting) for the efficient utilization by plants. Further, by applying organic manures at the time of field preparation or by raising and incorporation of green manures, sufficient quantity of nutrients can be supplied for plant growth. In addition, application of bio-fertilizers like *Azospirillum* and phosphobacteria, 5 kg each on 45th and 75th day after planting, by mixing it with FYM (500 kg/ha) or periodic application of *Amruthpani* along with irrigation would also improve the crop growth. The manures should be applied in the sides of furrows and incorporated into the soil while earthing up.

Several options are available for organic methods of supplementing soil nutrients with low cost. *Amruthpani* is a solution of 20 kg fresh cowdung, 1 kg jaggery, 1 kg gram flour, 200 ml sesame/gingelly oil, 5 litres of cow urine, 5 kg of bio fertilizers or oilcake and 500 gram ant hill soil or light soil. These ingredients are mixed in 100 litres water in a drum. The drum can be kept in a shaded place for 5 days. This quantity of liquid fertilizer is sufficient for one acre of sugarcane crop. The ingredients should be thoroughly mixed by a wooden stick twice daily. After 7 days the liquid is ready to be applied through irrigation water. Application of *Amruthpani* 4-5 times in a season boosts up crop growth.

Weed Management

A weed-free environment is absolutely essential for efficient intake of nutrients. This can be achieved by deep ploughing and removal of perennial weeds. Hand weeding and mechanical weeding at 30, 60 and 90 days after planting is better for long term benefits. Other appropriate measures to control the weeds should be practiced to minimize the production loss.

Earthing up

Normally, earthing up is done twice in sugarcane crop. Partial earthing up is done on the 75th day after planting, essentially to disturb the roots a bit and hence to trigger more tillers in the initial stage of the crop. This can be done by local *desi* plough or by lifting little soil from the side of root zone using a spade and spreading it across the row. Full earthing up is done around 120th day after planting. In this operation, soil from the ridge is thrown to both the sides of the plant towards furrows and these furrows will become ridges and *vice versa*. The newly-formed furrows will be later used for irrigation.

This full earthing up helps in preventing further production of tillers and provides sufficient anchorage to the crop against lodging. Earthing up of sugarcane plants helps in triggering new tillers, providing better aeration, covers and mixes applied fertilizers in the soil, better root development, checks growth of water shoots, provides sufficient anchorage and prevents lodging.

Detrashing and Propping

A normal growing cane stalk, on an average, bears 30-35 leaves under good growing conditions. But, for effective photosynthesis, only the top 8-10 leaves are sufficient. Most of the bottom leaves do not participate in the process and compete for the nutrients which otherwise could be used for stalk growth. It is important to remove the lower dry and green leaves during the 5th and 7th month and apply them as mulch in the interspaces. This facilitates a clean cultivation besides enhancing aeration. Movement inside the field becomes easier; disease pest incidence is reduced, easy to practise intercultural operations. The leaves can be used as mulching to prevent weed growth besides conserving moisture and ultimately decay into organic manure.

Propping means giving support to the canes to avoid lodging. Normally, this is done by tying the canes with one another using leaves. Sugar is synthesised in leaves; especially middle level green leaves contribute a lot in sugar production and thus the practice of propping by using those leaves to tie canes together should be avoided. It is advisable to use the dry bottom leaves for propping and to avoid young green leaves in the middle. Propping can be done in the 7th month, either by tying the canes in each row, or by bringing the canes of two rows together and tying them.

Intercropping

The wider spacing between the rows provides ample scope for intercropping within standing crop of sugarcane (Figure 13, 14, 15). Crops like green gram, cowpea, gram, potato, onion, wheat, coriander, lady's finger and melons can be effectively taken up as intercrop with sugarcane.

Different intercrops may be tried depending on location-specific factors. These intercrops help in reducing weed competition to the extent of 60 % in addition to effective utilization of land and give extra income to farmers. It is advisable to select nitrogen-fixing legume crops as intercrops, as they fix atmospheric N and improve the nutrient status of the soil upon incorporation after harvest. Intercrops also act as live mulch and preserve moisture and reduce the pest attack by being alternate hosts in some cases. Green manures raised as intercrop improve the soil fertility on incorporation.

Harvesting

Harvesting of sugarcane depends on sugar factory schedules. The crop is ready to harvest when the refractometer reading is around 17 – 19. Sucrose content in the plants will reach the most desirable level in the 10th month of the one year crop duration and canes will be ready for harvest within the next two months. While harvesting, care should be taken to cut the canes from the base, preferably 5 cm below the ground using axe or similar kind of implements. Improper harvest using sickles would result in the high sucrose- containing bottom part of the plant being left in the field itself, resulting in reduced cane harvest and less sugar yield. Harvesting using an axe is also preferable as there is no need of stubble shaving in the case of ratooning.

Figure 13. Intercropping (sugarcane + coriander).

Figure 14. Intercropping (sugarcane + fenugreek).

Figure 15. Intercropping (sugarcane + potato).

Table 3. Sugarcane cultivars suitable for SSI method of cultivation under different soil/climatic conditions

Soil/climatic conditions	Sugarcane cultivars
Drought resistant	Co 62715, Co 86032, Co 740, Co 87263, Co Or 04152
Heavy rainfall areas	Co 0671, Co 7508, Co 87263
Standing water/ low lying areas	Co 8021, Co Or 03151,
Cold tolerant	Co 62175
Saline soil	Co 6907
Acid soil	Co 6907, Co Or 03151
Red rot resistant	Co 87263, Co Or 03151, Co A 89085, Co Or 04152

Source: Department of Agriculture, Govt. of Odisha

SSI in Real Farming Situations: A Case Study

An on-farm trial was conducted for two consecutive cropping seasons of 2011-12 and 2012-13 using a participatory approach on cultivators' fields at Patuli Sahi village under Odagaon block of Nayagarh disrtrict in coastal climatic conditions of Odisha (India). In an interactive session, farmers of the village were informed about the objective of on-farm trials and 5 farmers mutually agreed to make their land available and participate in activities of the trial. The plot size was 500 m^2 for each individual farmer field for both the cropping seasons. SSI (Sustainable Sugarcane Initiative) technology was compared for cane yield advantage and economic returns with that of conventional method of planting at all the 5 locations during both the years.

In SSI technology, the bud chips were scooped out from upper 2/3 rd portion of healthy and disease free canes using a bud chipper and then after put in the nursery beds on 8th January, 2011 and 16th January, 2012. All the bud chips were put in a porous gunny bags and immersed in slurry of 2.5 kg *Trichoderma* culture and 10 liters of cow urine mixed in 50 liters of water for 30 minutes. Then the treated buds were taken out and shade dried. Bud chips @ 13000 /ha were taken for planting in all the fields of this study following the standard protocol of SSI technology. All the chipped buds sprouted after 6 days. Twenty-five days old seedlings were transplanted in the main field on 2nd February, 2011 and 10th February, 2012 at a spacing of 120 X 60 cm distance. A mid late maturing (12 months) sugarcane variety - Co Or 04152 (Raghunath) was used in this study. The recommended fertilizers doses were 250: 100: 60 kg N, P_2O_5 and K_2O/ha. During final land preparation, Chlorpyriphos 50 % EC was applied to the main field @ 2 litres/ha to control the incidence of termite and early shoot borer. Before transplanting of the sugarcane seedlings, well decomposed FYM @ 20 tonnes/ha was mixed up with 1/3rd of total N, full dose of phosphorous and half of K and placed in the furrows followed by a light irrigation. One pre emergence application of Atrazine @ 2 kg a.i./ha at 2 DAP (days after planting) and two hand weeding at 60 and 90 DAP were done to control weeds. Remaining 2/3 rd dose of N was applied as top dressing in two equal splits, one at 45 DAP and rest one with remaining dose of K at 90 DAP along the furrows after weeding and hoeing. At 75 DAP, *Azotobacter* and phosphorus solubilising bacteria each at 5 kg /ha were mixed with 250 kg of well decomposed FYM and applied to the field. Irrigation was given in alternate furrows as and when required to keep the field moist except in rainy season. The crop was harvested on 8th February, 2012 and 21st January, 2013. All agronomical packages of practices were followed to raise the crop in both the techniques of planting. Observations on sugarcane growth and yield attributes were recorded at the appropriate stages and compared accordingly after working out economics.

Figure 16. Sugarcane crops raised through SSI technology and conventional methods.

Table 4. Yield attributes and yield of sugarcane grown through SSI technology and conventional 3-bud setts planting method in Nayagarh district of Odisha (Mean of two years)

Planting technology	Percentage survival/ germination	No of millable canes/ clump	Length of cane (cm)	Girth of cane (cm)	NMC' 000/ha	Single cane wt (kg)	Cane yield (t/ha)	Percent increase in cane yield over conventional method
SSI	88	9.6	205.20	3.1	116.13	1.12	105.00	18
Conventional 3-bud setts	55.81	5.2	197.1	2.8	92.66	1.04	89.00	-

Percentage of seedling survival in SSI technology and percentage bud germination in conventional method were compared and presented in Table 4. The results exhibited that the seedling survival rate was 88 % in SSI technology as compared to only 55.81 % bud germination in conventional method, creating a huge difference in plant population which is ultimately responsible for higher number of millable canes in SSI technology as compared to conventional method. Moreover, there was scope for gap filling with nursery raised seedlings in SSI technology which helped to ensure the initial plant stand in the field. The crops raised through SSI technology and conventional method are shown in Figure 16.

Table 5. Economics of sugarcane cultivation (RS/ha) in SSI technology vs. conventional method of sugarcane planting

Sl No	Particulars of sugarcane cultivation	SSI technology of sugarcane planting			Conventional method of sugarcane planting		
		Units	Unit cost (Rs.)	Total (Rs.)	Units	Unit cost (Rs.)	Total (Rs.)
1	Land preparation (ploughing days)	12	500	6000	12	500	6000
2	FYM (ton)	20	1200	24000	20	1200	24000
3	Seed material (ton)	0	0	0	10	2250	22500
4	Hiring charges of nursery trays & bud chipper	25	140	3500	0	0	0
5	Cost of coco pith/coir dust (in bags)	10	80	800	0	0	0
6	Bud chipping (in man days)	10	200	2000	0	0	0
7	Bud treatment/sett treatment (in man days)	3	200	600	3	200	600
8	Opening of furrows (ploughing days)	3	500	1500	4	500	2000
9	Planting(in man days)	15	200	3000	10	200	2000
10	Bio fertilizers (kg)	10	40	400	0	0	0
11	Plant protection measures	1	1500	1500	1	1500	1500
12	Chemical fertilizers (N :P_2O_5 : K_2O : 250:100:60 kg/ha)	-	-	10000	-	-	10000
13	Weedicide (Atrazine) (kg)	2	75	150	0	0	0
13	Intercultural operations 2 times (in man days)	120	200	24000	120	200	24000
14	Detrashing, wrapping & propping (in man days)	100	200	20000	108	200	21600
15	Harvesting, cleaning & transportation by head load up to truck(in man days)	230	200	46000	238	200	47600
16	Irrigation	17	500	8500	15	500	7500
	Total cost of cultivation (Rs.)		0	151950		0	169300
	Gross return (Rs.)	105	2250	236250	89	2250	200250
	Net Return (Rs.)	-	-	84300	-	-	30950

Source: On-farm study in farmers' field from Nayagarh district of Odisha.

Average number of millable canes/clump was 9.6 in SSI technology as compared to 5.2 in conventional method. Length and girth of canes in SSI technology were 205.2 and 3.1 cm, respectively as compared to 187.1 and 2.4 cm in conventional method of planting. Similarly, number of millable canes were also higher (116.13'000/ha) in SSI technology as compared to conventionally grown crop (92.66'000/ha) which clearly endorses the result of higher number of millable canes/clump as discussed above. Average cane weight was higher (1.12 kg) in SSI technology as compared to that obtained under conventional method (1.04 kg) of planting. Higher plant stand along with higher yield attributing characters resulted in higher cane yield of 105.0 t/ha in SSI technology as compared to 89.0 t/ha under traditional three bud setts planting.

CONCLUSION

In SSI method the seed cost is reduced up to 90% as compared to conventional method and the entire length of cane can be used for extraction of the buds to be used as planting material. The plant mortality rate is reduced as the seedlings are graded before transplanting. The length and weight of individual canes increase due to less competition for sunlight, nutrition and water. Consequently, more number of millable canes/clump are also be obtained. It is easy to transport the planting materials/buds to longer distance. The intercultural operations are also convenient due to wider spacing. The cane yield obtained under SSI technology was higher by 18 to 20 % over conventional method of planting. Thus, it may be suggested that the SSI technology of sugarcane planting is worth adopting particularly by the small and marginal farmers since it is not only high yielding, cost effective and sustainable but also attracts a large number of farm women due to easy planting operations involved in SSI technology.

REFERENCES

[1] Naik Ravindra, Annamalai SJK, Vijayan Nair N and Rajendra Prasad N (2013). Studies on Mechanization of Planting of Sugarcane Bud Chip Settlings in Protrays. *Sugar Tech.* 15(1): 27-35.

[2] Srivastava KK, Narismhan R and Shukla RK (1981). A new technique for sugarcane planting. *Indian Farming* 31(3): 15-17.

[3] Mohanty M and Nayak PK (2011). Economizing seed cane quantity by reducing sett size and bud number with sett treatment in sugarcane cultivation. *Indian Journal of Sugarcane Technology* 26(2): 59-60.

In: Current Status of Sugarcane Research in India
Editors: A. K. Tiwari, M. Lal and A. K. Singh
ISBN: 978-1-63463-458-8

Chapter 11

A Century of Sugarcane Red Rot Research in India

Sangeetha Panicker** and R. Velazhahan
Tamil Nadu Agricultural University
Tamil Nadu, India

Abstract

This chapter looks back at about hundred years of research on red rot of sugarcane in India. Aspects like epidemics caused due to red rot of sugarcane in India and the associated quantitative and qualitative losses have been highlighted. Some of the basic and advanced researches on the pathogen morphology and molecular characterization of *Colletotrichum falcatum*, diagnostic methods and the induction of systemic resistance against the pathogen have been described.

Keywords: Sugarcane, red rot, *Colletotrichum falcatum*, India

Introduction

History of Sugarcane

Sugarcane agriculture in India dates back to the Vedic period. Gur, a name for raw sugar, has originated from the word "Gaura" a well known dynasty that ruled Bengal in 3000 BC. Later the word "sugar" is said to have originated from the Sanskrit word "Sarkara" as mentioned in Sanskrit literature 1500 – 500 BC. Chinese writing of 800 BC also quotes India as the origin of sugarcane. There is also a mention about Sweet cane sugar in the old testament in both Isaiah 43:24 and Jeremiah 6:20. Myths and beliefs of sugarcane are many. Sugarcane in India is considered spiritual and holy and is said to signify prosperity and well being. Sugarcane is used at the occasions of several festivals like Pongal in Tamil Nadu,

* Email for correspondence: sangeetha_murali@hotmail.com.

Sankraanti in the Northern states and Lohdi in Punjab. In Atharva Veda, there is a reference of sugarcane being used as a magical ingredient. In Hindu religious stories, Kama Deva, the God of love is said to hold a bow of sugarcane meaning that love is as sweet as sugarcane. So also in Goddess Devi is said to hold a bow of sugarcane of red variety in her hand and also Lord Ganesha is referred to hold sugarcane.

Journey of Sugarcane from India

There are different thoughts regarding the journey of sugarcane around the World probably because of the lack of proper literature and writings in the early days. From India sugarcane travelled to China in 250 BC. Darius the Great brought sugarcane to Persia from India around 500 BC. Later, Saracens introduced sugar to Egypt, Sicily and Spain and by 1 AD it reached Java or Indonesia. Later when “Alexander The Great” invaded India in 327 BC, he was fascinated by the crop sugarcane and called it “Honey yielding reed” and carried it Westward. By 8th century it spread to Italy, France and Spain. Marco Polo introduced sugarcane to Venetians in 12th century. In 1493 Christopher Colombus spread it to America. By 15th century it reached Europe via Egypt. In1498, Vasco De Gama introduced sugarcane to Portugal and Lisbon. Today sugarcane has established itself as an important cash crop around the world.

History of Red Rot Disease

As early as 8th – 7th century BC, the disease was mentioned in Vedic text as Aitereya Aryanaka. In this text it was also mentioned about Indian epic Mahabaratha that there was a Pundra kingdom where the people suffered from a skin disease called Pandu and it is said the name Pandu originated from a sugarcane variety then found in that region which was affected by a disease of red patches. One also finds the mention of red rot in the Buddhist literature [1]. The disease was so prevalent in UP and Bihar where Gautama Buddha started preaching after enlightenment. Buddha referred to red rot of sugarcane as Manjitthika after a red dye. He quoted “just as the disease known as Manjitthika falls on a field of ripened sugarcane, that field does not last long”.

Being one of the oldest diseases of sugarcane, red rot is distributed worldwide. This disease has been a major constraint to sugarcane productivity for more than 100 years in India. Several important commercial varieties have been wiped out of cultivation due to this disease from time to time. The pathogen causes severe losses in yield and quality of the crop in the Indian Sub continent [2, 3].

In India, the first red rot epidemics occurred in the Godavari delta in 1895. The area under sugarcane reduced from about 2500 ha to 500 ha due to this disease by 1899. Hence, the government deputed Mr. C.A. Barber to study the crop and the disease. In his report, he described the causal agent of red rot of sugarcane as *Colletotrichum falcatum* Went. [4]. Later, ‘Sugarcane Breeding Institute’ was established at Coimbatore in the year 1912 under the leadership of Mr. C.A. Barber and T.S. Venkatraman to undertake researches on sugarcane development and cultivation practices. Venkatraman was the first scientist in the world to use wild species for improvement of cultivated crop. After about two decades, Tamil

Nadu Agricultural University established a research station in Tamil Nadu to intensify researches on Sugarcane in 1935. In the Pre 1970s, the red rot disease was largely confined to the subtropical regions of India but it was prevalent in other states of the country viz., Uttar Pradesh, Punjab, Bihar, Orissa, West Bengal and Rajasthan. Later, the disease spread to the tropical regions such as Andhra Pradesh, Tamil Nadu, Pondicherry and Kerala [5].

Dr. Butler published a series of papers on this disease including a landmark research paper on sugarcane diseases of Bengal. These papers gave a detailed description of red rot of sugarcane and its epidemiology [6, 7]. Two different forms of *C. falcatum* (light and dark forms) were reported by Ramakrishnan [8]. Studies by Atkinson [9] showed that *C. falcatum* even produced stromata bearing setae. Later, Chona et al. [10] identified an Indian strain of the pathogen producing Stroma. In the year 1938, there was another epidemics of red rot in the variety Co 213 which shook the entire country especially UP and Bihar and reduced sugar production by nearly 70,000 tonnes. This led to deeper research on red rot problem and a series of publication on the subject [11, 12]. This further led to the establishment of a Central Institute entirely for sugarcane research in Lucknow in 1952 (now known as IISR-Lucknow).

Early Epidemics of Red Rot in India

Epidemics of red rot has been common since 1895 (Table 1). The disease affected *Sacchrum officinarum and S. barberi* clones and later many newly released commercial varieties succumbed to the disease. Several important genotypes including Co 213, Co 281, Co 290, Co 312, Co 313, Co 419, Co 421, Co 453, Co 527, Co 658, Co 975, Co 997, Co 1148, Co 7717, CoC 671, CoC 85061, CoC 86062, CoC 92061, CoJ 64, CoJ 82, CoJ 84, CoLK 8001, CoLK 8102, CoS 245, CoS 510, CoS 770, CoS 767, CoS 802, CoS 8436, CoSe 93232, CoSe 92423, BO 3, BO 10, BO 11, BO 14, BO 17, BO 29, BO 34, BO 54, BO 120, BO 128, etc. were wiped out of cultivation due to red rod. The major disease epidemics occurred in India over the decades have been summarized in Table 1 [13].

Economic Importance of Red Rot

Ever since the initial occurrence of red rot in 1890, the disease has caused great loss worldwide. Unlike other diseases (e.g. yellow leaf disease or rust), this pathogen directly attacks the economically valuable stalk tissues and hence even limited infection can bring about drastic change in juice quality. Severe epidemics of red rot occurred in India since 1895 [13,14], it affected several excellent varieties and hardly a few genotypes were unaffected. Many genotypes had to be withdrawn from cultivation. In Tamil Nadu variety like CoC 92061 was withdrawn immediately after its multiplication and promotion due to red rot.

Red rot affects sugarcane crop at three stages. Firstly, if affected at nursery stage, there is a loss in germination itself or seedling death occurs. Nodal infection is reported to cause 15-20% loss in germination [15, 16]. Dormant infection in seed cane may cause upto 73% loss in germination [17]. Secondly, red rot infection causes loss in cane yield. Chona [18] reported that in India, 2/3rd of the cane is lost due to severe red rot infection. Kumar et al. [19] recorded reduction in length, girth and weight of cane due to red rot infection. More than 50% loss in cane yield has been reported in India due to red rot in varieties Co 6304, CoC 671,

CoC 85061, CoC 86062 etc and yield loss upto 100% was recorded in different factory regions [20, 21, 22]. Thirdly, there is loss in quality of the juice.

Table 1. The major red rot epidemics on sugarcane in India

Year	Area	Major genotype affected
1895-1901	Godavari Delta (Andhra Pradesh)	*Namalu Keli*, Bourbon, Ashy Mauritius, Striped Mauritius
1902-1913	Champaran, Muzaffarpur (Bihar)	Bourbon, Striped Mauritius
1922	Jammu (Kashmir)	
1925	Punjab	
1932	Pusa (Bihar)	Co 210
1936	Inthiathoke (Gonda, U.P.)	
1937	Gola, Nagina (U.P.)	
1938-40	Uttar Pradesh, Bihar	Co 213
1946-47, 1951-68	Uttar Pradesh, Bihar	Co312, Co 313, Co 419, Co 421, Co 443, Co 453 Co 513, Co 997, CoS 510, CoS 514, BO 3, BO 11, BO 17, Mamchuah (local variety)
1960-70	Andhra Pradesh	Co 421, Co 419, Co 997, Co 62175
1971-75	Tamil Nadu	Co 419, Co 658, Co 6304
1978-83	Andhra Pradesh	CoC 671, CoA 7701
1981-85	Kerala	Co 997, Co 419, Co 785
1970-85	Uttar Pradesh, Haryana, Punjab, Bihar	Co1148, Co 7717, CoJ 64, Co S 527, CoS 659, CoS 770, BO 29, BO 32, BO 54, BO 70
1992-97	Gujarat , Tamil Nadu	CoC 671, CoC 85061, CoC 86062, CoC 90063, CoC 92061
1991-95	Uttar Pradesh	CoS 767, CoS 802
1990-99	Andhra Pradesh	Co 7508, CoC 86062
2000-2005	Uttar Pradesh	CoS 8436, CoSe 92232, CoSe 92423, BO 120, BO 128

Source- Duttamajumder [13].

Table 2. Juice quality in healthy and diseased canes

Variety	Juice Quality Parameters								
	Brix (%)			Polarity (%)			Extraction %		
	Healthy canes	Diseased canes	Percent Reduction	Healthy canes	Diseased canes	Percent Reduction	Healthy canes	Diseased canes	Percent Reduction
Co 7314	17.08	15.88	7.03	13.72	12.70	7.43	72.4	68.8	4.9
Co 1148	16.58	13.50	18.58	13.31	10.63	20.14	74.0	59.2	20.0
CoS 802	18.43	15.78	14.38	15.42	12.58	18.42	68.0	60.4	11.2
CoJ 64	18.68	15.50	17.02	15.47	11.89	13.14	74.0	61.2	17.3
CoL 9	19.83	13.38	32.53	16.62	10.62	36.10	76.0	45.0	40.8

Source: Viswanathan [14].

The pathogen impairs sucrose metabolism and reduces total carbohydrate in the diseased cane especially in susceptible varieties [23]. Changes in sucrose and glucose content of the juice due to red rot infection was first observed by Went who found an increase in glucose content and a decrease in sucrose content due to inversion of sucrose.

Later, Butler confirmed this observation and reported that the decrease in sucrose content was due to inversion and not due to consumption by the pathogen. Moreover, there was an increase in the total soluble salts, acidity, reducing sugars and gum along with a decrease in pH of the juice [2].

It was also observed that juice obtained from red rot infected canes, when used for sugar making, did not crystallize well, hence affecting sugar recovery [13]. Infection also reduces brix of juice [24,25]. Satyavar et al. [26] reported that red rot infection reduced extraction by 7.1 to 32.5 %, polarity by 7.4 to 38.7 % and purity co-efficient by 7.1 to 32.5 % and there was an increase in reducing sugar by 19.2 to 40.95 %. The percent reduction in Brix, polarity, juice extraction etc due to red rot infection in some varieties are given in Table 2.

It was also found that in red rot infected canes, there was slight increase in lipids, proteins, calcium and iron content and an increase in activity of enzymes like amylase, protease, peroxidase, invertase, beta glucosidase, and catalase [27]. During the milling process mixing of juice from healthy and diseased cane resulted in spoilage of entire juice due to inversion of sucrose. Similarly, jaggery setting was also affected due to red rot infection.

The worst crisis faced by sugar industries in India especially Tamil Nadu was from 1896 to 2002. During this period the western parts of Tamil Nadu experienced severe epiphytotics of red rot. Introduction of red rot through CoC 671 and CoC 92061 led to spread of red rot in a severe manner in the western regions of the state which escaped red rot during 1970's and 1980's [20,21,14]. It was found that, the introduction of the highly susceptible variety (CoC 92061) was the major reason for devastating the entire cane industry. Finally the sugar industry in the state revived with the introduction of Co 86032, which has withstood the onslaught of red rot for more than 15 years and still continues.

STUDY OF THE FUNGUS IN INDIA

Morphological Characterization of *C. Falcatum* Isolates

Earliest study in the world on the fungus was carried out by Went in 1893. He studied in details the symptoms, proved the parasitism and named its imperfect stage as *Colletotrichum falcatum* Went. and carried out studies on its life history [28]. The perfect stage *Physalospora tucumanensis* Speg. was first recorded by Spegazzini from Afghanistan in 1896 [29]. Later, in 1954, Von Arx and Muller from Germany transferred this fungus to the genus *Glomerella tucumanensis* (Speg) [30].

In India Butler and Khan [31] studied the similarities between *Colletotrichum lineola corda* of *Sorghum vulgare* and *Colletotrichum falcatum* in sugarcane and concluded that though the cultures and symptoms looked similar, the causal organisms of sugarcane and cereal were distinct. Ramakrishnan [8] studied the physiological aspects of growth and behavior of *C. falcatum* and found that *C. falcatum* grows well in a number of standard media including French bean and Richards agar and found oatmeal agar to be the best medium. He found that pH 4.5 to 5.0 was optimal for growth and spores formed at 15° C and 30° C were of normal size (19.9 to 27.6 μm). The thermal death point of conidia was found to be five minutes at 51° C. As regard to the source of carbon and nitrogen, sucrose was found to be the best source of carbon while, aspargine and potassium nitrate were found to be the best source of nitrogen. The optimal carbon: nitrogen ratio was 5:1. He reported dark and light type of strains which differed in their pathogenicity and sporulation capacity. Later, Chona and Srivastava [32] reported the occurrence of perfect stage in culture at Indian Agricultural Research Institute, New Delhi. They found that isolates of light type produced the perfect stage more commonly than the dark isolates. Chona and Bajaj [33] reported its occurrence in nature on leaf lamina, midrib and leaf sheath and dried foliage. The fungus was capable of growing in soil and producing acervuli [18, 34, 35, 36,]. However, the workers in India were initially not able to detect the perfect stage of the fungus under field conditions and were doubtful about its presence in India. Later in 1996, Duttamajumder was able to detect the presence of the perfect stage in India in sufficient numbers. He also reported the presence of perithecia.

Conidial Characteristics

Conidia are produced on short conidiophores, closely packed inside the acervulus. Conidia are single celled mostly falcate, but some are straight, muticate or slightly punctuate, transparent to densely granular and frequently guttulate. Sometimes empty black pseudopycnidia are produced. In general, length and width of conidia are highly variable. Several studies have been made in India regarding size of conidia. Varied opinion has been given regarding the size of conidia. Chona [18] studied two isolates and reported the size to be 27.1 X 4.99 µm and 20.3 X 4.92 µm. Singh and Rana [37] studied the morphological characters of three isolates and reported the length and breadth to be 26.6 X 4.12 µm, 23.2 X 5.0 µm and 22 X 6.1µm respectively. Gupta et al. [38] studied the cultural characteristics of a new biotype R183 and reported that its conidial length varied from 33 to 37.4 µm and width varied from 4.4 to 4.95 µm. Agnihotri [23] reported that the sickle shaped or falcate conidia measured 16-40µm X 5 to 7 µm in size and contained oil globule in the centre. Conidia develops in pink or salmon coloured water soluble mucilagenous mass and when produced rapidly the upper portion of the acervuli gets covered with shining droplets. Numerous black setae develop in and around the stroma, these are 100-220 µm in length and separate with a bulbous base. Jothi [39] reported that all the isolates studied at Coimbatore in Tamil Nadu State of India produced acervuli in culture whose diameter ranged from 0.639 to 1.54 µmm. Setae were also found in all the isolates whose number ranged from 3 to 20 per acervulus and the length ranged from 90 – 220 µmm, conidiophores length ranged from 120 to 330 µmm. Later, Duttamajumder [40,13] described the characteristic of *C. falcatum* as follows, the colony was grayish white with sparse aerial mycelium, the conidial masses were salmon pink colored in case of light races, while some cultures produced abundant grayish white aerial mycelium with poor sporulation and no distinct acervuli. Both dark and light races do not produce sclerotia, setae sparse, conidia falcate, fusoid, apices obtuse, 15.5(25-26.5) 48 µmm X4(5-6)8 µmm and content are granular and sometimes contain oil globules. Conidia do not have resting period and germinate immediately after production. Usually it follows sub polar germination from one or both ends of the conidium. After germination the hyphae may immediately produce appresoria or may continue to grow. From the appresorium infection pegs develop which enters the host tissue and initiate infection process. Under stress condition, sometimes the conidia germinate and produce a fussion aggregate to overcome the stress situation [41]. Many strains of *C falcatum* in old cultures and diseased stalks produce round double walled structures called chlamydospores that remains dormant in the soil [6]. The chlamydospores serve dual purpose of close adhesion to the surface of host plant and accumulation of enzymatic energy to secure penetration of its wall [31]. Agnihotri [23] induced chlamydospores formation in PDA by sprinkling 50-100 mg of natural soil over 10 days old colony of *C. falcatum*.

Certain isolates of *C. falcatum* also produce black hard structures of stromatic bodies which consists of fertile hyphae, some of which perform true function as conidiophores and bear typical falcate conidia. The key morphological identification feature of the fungus is the asexual fruiting body or acervulus, in which dark brown seatae develop. The setae are septate stout structures with bulbous base, tapering towards the tip. The setae are usually sterile structures, however it has been reported that under favorable conditions when tip remain hyaline, it can contribute to production of conidia [40].

The perfect stage of the fungus namely *Glomerella tucumanensis* is an ascomycetes fungus and produces ascopspores in typical eight spored ascus in perithecium. Perithecia is inconspicuous and entirely embedded in cane tissue. Asci are clavate measuring 70-90 μm X 75(13)-18 μm, hyaline straight to slightly fusoid, with one celled ascospores arranged in a biserriate pattern in the ascus [33, 40, 13].

Chona and Srivatava [32] reported that isolates of *C. falcatum* are usually unstable in culture and there are no correlation between growth rate, morphology, physiological behaviour and virulence. Malathi et al. [42] examined variation in cultural, pathogenic and genetic characters using nine major pathotypes of *C. falcatum* from tropical and sub- tropical regions of India which are mainly used for scoring of red rot in sugarcane. They reported that culturally, the isolates exhibited variation in mycelial growth, colour, texture, acervuli initiation, sporulation and conidial germination. Of these, acervuli initiation and sporulation had a relation with the virulence of the pathotypes.

Pathotypes in *C. Falcatum*

Earlier studies revealed greater virulence of pathotypes from tropical India over the sub-tropical pathotypes [43, 44]. It is well known that pathogen diversity can determine the dynamics of epidemics. Red rot pathogen shows a great diversity in virulence as a number of pathotypes are known to occur in nature which has been classified on the basis of host differential reaction [45]. As the host reaction is influenced by many climatic factors like temperature, humidity, time of inoculation, age of culture etc., the results sometimes become very confusing [46]. Moreover sugarcane being a crop of 10-12 months restricts the use of many *C. falcatum* isolates to be tested under field conditions.

Alexander et al. [47] reported different pathotypes of *C. falcatum* isolates in India on the basis of their differential reaction. Subsequent workers reported the existence of different pathotypes of *C. falcatum* using host differentials [48, 49, 50]. On the basis of these host differentials, Satyavir [51] summarized Cf 01 (Co1148), Cf 02(Co7717), Cf 03 (CoJ64), Cf 07 (CoJ64), Cf 08 (CoJ64/CoJ84), Cf 09 (CoS767) from the North West Zone and Cf 04 (Co419), Cf 05 (Co997), Cf 06 (CoC671) and Cf 10 (35A261) pathotypes from East Coast Zone. Subsequent studies on pathogenic variability during 1993-2000 revealed the existence of four new pathotypes *viz.,* Cf 07 (CoJ64), Cf 08 (CoJ64), Cf 09 (CoS767), and Cf 10. But all the pathotypes except Cf 09 were avirulent on CoS767. The breakdown of resistance in this cultivar was noticed in Haryana and Uttar Pradesh in recent years. These studies confirmed the appearance of a new pathotype (Cf 09) capable of breakdown of resistance of this widely cultivated cultivar in North West Zone [52].

Malathi et al. [53] reported that pathogenicity of both the pathotypes was influenced by their respective/host specific parental cultivars. These pathotypes were well differentiated earlier on the basis of pathogenicity, serological and molecular studies [54]. Although variations in pathogenic, serologic, molecular and cultural characters are known [44], origin of new pathotypes or adaptation of *C. falcatum* to a sugarcane cultivar which was hitherto resistant is not clearly understood.

Recent Advances on Red Rot Research in India

With the development of science and technology, plant pathological research is also changing. Any plant disease has to be first diagnosed and then managed or controlled. In the traditional method diseases are often diagnosed only after the occurrence of visible symptoms and this is possible only after major damage has already been done. Moreover in traditional methods, identification is based on lab studies by subject experts and finally controlled by use of chemicals. Biological control methods followed the traditional methods of disease management but today biotechnology, nanotechnology, computer technology, bioinformatics etc have revolutionized the field of plant pathological research. Biotechnological approaches aim at diagnosis of disease at molecular level by using PCR, technique which is more sensitive and accurate. In biotechnology the principle of the complex immune response in plants and pathogens and their interaction is also being exploited. Nanotechnology plays an important role in nanogenetic manipulation of plants to develop disease resistant plants and disease control by controlled delivery of functional molecule etc. Computer technology plays a major role in red rot prediction models [55], epidemiological studies, visual evaluation and image analysis etc. Finally bioinformatics provide huge quantity of information in any biological fields like genome sequence from disease causing organisms that helps in understanding plant pathogen interaction or the genome sequence from wild plants help in identifying resistant genes against pathogens etc.

Diagnosis of Red Rot

In the case of sugarcane the management of red rot disease by using disease-free seed canes for planting is impractical due to the difficulty in diagnosing the dormant infections of the fungus in seed canes under field conditions [56].

Early identification and control is important to avoid severe losses, though personal consultation with a specialist is preferable, it is not always feasible due to limitations of time, cost, availability of the expert etc. However, with the availability of computers and the Internet this requirement can be easily met through suitable software tools e.g., “The Sugarcane Doctor” [57]. The development of molecular techniques for identifying and distinguishing sugarcane pathogens also continues to make rapid progress [58].

Serological Diagnosis

The use of Enzyme Linked Immunosorbent Assay (ELISA) technique for detection of red rot pathogen in the infected host was found to be useful in the early diagnosis of cane infection. Polyclonal antisera were raised against the unfractioned protein of *C. falcatum,* a partially purified 101 kDa protein and the serological techniques were standardized to detect the pathogen in ELISA, DIBA (dot immunobinding assay) and Western blot [59]. Hiremath and Naik [60] reported the DIBA technique as a simple, rapid and specific for laboratory diagnosis of sugarcane red rot infection in the planting material at an early growth stage. Several scientists have developed different serological and molecular methods for detecting presence of *C. falcatum* in sugarcane. Viswanathan et al. [59] developed polyclonal antisera against unfractionated protein of *C falcatum* which were effective in detecting *C falcatum* by ELISA, DIBA and Western blot method. Khalid et al. [61] developed polyclonal antibodies that were highly specific to *C. falcatum* and very effective in ELISA method. Other methods include direct antigen coating enzyme linked immunosorbent assay, DIBA [62] and also by using SCAR marker [63]. Molecular and pathological characterization of *C. falcatum* was carried out by Narendran et al. [64], who used three different marker systems to characterize 25 isolates of *C falcatum* for North Eastern states of India and assessed the pathogen diversity. He found that isolates Cf 01, Cf 08 and RR 15 were the most virulent and Cf 07 the least virulent and these 25 isolates were classified into six clusters. Similar studies on *C. falcatum* isolates from North West zone of India was reported by Saksena et al. [65], who identified the presence of two new pathotypes.

Molecular Characterization of *C. Falcatum* Isolates

Identification of *C. falcatum* became complicated by traditional methods because of overlapping range of conidial morphology and variation in colony characteristics. The criteria followed in traditional methods were not adequate to differentiate species because of variation in morphology and phenotype among species under different environmental conditions. The application of molecular markers to fungal taxonomy promises to clarify the genetic relationships of phytopathogenic fungal groups [66]. This is especially important in fungal taxa such as *Colletotrichum*, which are not clearly distinguished by their morphology. Identification through genetic characterization is important because members of the genus *Colletotrichum* cause disease on virtually every significant agricultural crop worldwide. A variety of molecular approaches have been developed and used, since morpho-taxonomic criteria are not accurate to discriminate between the *Colletotrichum* species.

Some of the molecular markers that can be used are Restriction Fragment Length Polymorphism (RFLP) [67, 68]. Random Amplified Polymorphic DNA (RAPD) [69], Arbitrarily primed (ap)-PCR [70], Amplified Fragment Length Polymorphism (AFLP) [71], Simple Sequence Repeats (SSRs or Microsatellites), Inter Simple Sequence Repeats (ISSR) [72] and Internal Transcribed Spacer (ITS) [73, 74] etc. These markers detect polymorphism by assaying the subsets of the total amount of DNA variation in a genome.

Sreenivasaprasad et al. [74] performed DNA sequence analyses, to characterize and resolve the taxonomic complexity of *Colletotrichum* sp. Madan et al. [75] analyzed the molecular variability of five different *C. falcatum* isolates (Cf 01, Cf 02, Cf 03, Cf 04 and Cf

05) by using six RAPD primers. It separated the pathogenic variants Cf 01 and Cf 02 in one cluster and Cf 03 & Cf 04 in the other cluster. Cf 05 appeared to be related to both of these clusters. Mohanraj et al. [54] used 61 random primers in RAPD studies and found the presence of four groups among the isolates of *C. falcatum* i.e. Gr-1 (Co 7717); Gr-2 (Co 1148); Gr-3 (CoC 671, CoC 92061, CoC 85061), and Gr-4 (CoC 90063, CoS 767). Suman et al. [76] also examined six isolates using 40 RAPD primers and reported 2 UPGMA clusters of isolates. Cluster I included pathotypes Cf 01 & Cf 09, which were isolated from altogether different hosts. Cluster II included rest of the four pathotypes and amongst them the highest similarity (0.962) was observed between Cf 02 and Cf 08. On the basis of origin, the expectation of high similarity was between Cf 03, Cf 07 and Cf 08. Alvi et al. [77] studied the DNA based genetic variation for red rot resistance in sugarcane by using 300 RAPD markers.

Mishra and Behera [78] studied the pathogenic and molecular variability of *C. falcatum* isolates collected from Andhra Pradesh and Orissa following Restricted fragment length polymorphism (RFLP) analysis of internal transcribed spacer region of ribosomal DNA (rDNA) of Cf 89V74, Cf 671 and Cf Vittal (*C. falcatum* isolate from Vittal, Karnataka) by four restriction enzymes, *Alu* I, *Msp* I, *Rsa* I, and *Pvu* II. The results revealed two distinct groups *viz.*, Group 1-Cf 89V74 and Cf Vittal; Group 2-Cf 671. The molecular and pathological characterization of *C. falcatum* infecting subtropical Indian sugarcane by using three different markers *viz.*, RAPD, Universal Rice Primers (URP) and Inter Simple Sequence Repeat markers classified 25 isolates into six clusters at 34% genetic similarity and the isolate Co Pant 84212 was found to be genetically diverse [79].

Wijesekara et al. [80] examined 20 *Colletotrichum* isolates from 14 different crops including sugarcane using 16 primers in RAPD technique and reported that sugarcane isolates 4800 and 4803 produced an identical banding pattern while difference existed in their morphological characters. This phenotypic identification is time consuming, expertise - specific and not always fully discriminative.

Induction of Defense Mechanisms

Induced resistance is defined as an enhancement of the plant's defensive capacity against a broad spectrum of pathogens and pests that is acquired after appropriate stimulation. The resulting elevated resistance due to an inducing agent upon infection by a pathogen is called induced systemic resistance (ISR) or systemic acquired resistance (SAR) [81]. The induction of systemic resistance by rhizobacteria is referred as ISR, whereas that by other agencies is called SAR [82]. PGPR are root colonizing bacteria with beneficial effects including plant growth promotion and biological disease control. In recent years, the use of PGPR as an inducer of systemic resistance in crop plants against different pathogens has been demonstrated under field conditions [83, 84, 85]. The utilization of natural PGPR strains as inducers of plant defense responses may increase the chance of their applicability and offer a practical way to deliver immunization.

Resistance mechanisms attain their maximum effectiveness at four to five days after the application of an inducing agent, but the level of persistence of resistance generally decreases over time. These criteria determine the number of applications of PGPR needed to maintain the resistance level in the crop plants. In sugarcane, induction of resistance by PGPR persists for 90 days of crop growth [84]. Generally, the durability of resistance by PGPR differs from crop to crop and also due to different bacterial strains. In sugarcane, application of PGPR as sett-treatment induced systemic resistance against *C. falcatum* in addition to enhanced sett germination, tillering and growth of the cane both under controlled conditions as well as field conditions.

Ramesh Sundar et al. [86] studied the Induction of systemic acquired resistance (SAR) by pre-treatment with synthetic signal molecules CGA-245704; Benzo (1, 2, 3) thiadiazole - 7 - carbonic acid S - methyl ester (BTH) and salicylic acid (SA) induced the phenylalanine ammonia-lyase (PAL), peroxidase (POX), polyphenol oxidase (PPO) and accumulation of phenolics in systemically protected sugarcane stalks. The study clearly established that systemic acquired resistance holds a promise in managing red rot in elite commercial varieties under field conditions and it can be used as an effective management strategy for control of the disease in an environment expected to favour a disease outbreak.

Ramesh Sundar et al. [87] reported the application of synthetic signal molecule, Acibenzolar-S-Methyl (CGA-245704) as a soil drench or along with rooting mixture induced resistance in sugarcane to challenge inoculation with *C. falcatum*. An induced systemic resistance effect was found to persist up to 30 days in the pre treated cut canes and increased phenolic content and accumulation of pathogenesis-related (PR) proteins, *viz.*, chitinase, β 1,3-glucanase and thaumatin-like protein (PR-5), were observed in the treated canes compared to untreated control. Ramesh Sundar et al. [88] reported plant activators capable of inducing systemic resistance in sugarcane. The plants pretreated with synthetic signal inducers (Acibenzolar S-methyl, ASM) restricting the pathogen colonization inside the inoculated cane stalk tissues which confer a high degree of resistance to *C. falcatum.*

Viswanathan and Samiyappan [89] reported the induction of chitinase in sugarcane in response to red rot pathogen infection or saprophytic pseudomonads treatment. In *Pseudomonas* mediated induced systemic resistance in sugarcane against red rot disease, induction of β-1,3-glucanases, chitinases and thaumatin-like proteins (TLPs) has been reported by Viswanathan et al. [90]. These authors have also shown strong antifungal activities of purified sugarcane chitinases against *C. falcatum.*

Viswanathan et al. [91] assayed the induction of chitinases and thaumatin-like proteins (TLPs) at different time intervals in sugarcane varieties differing in resistance to *C falcatum* after pathogen inoculation. The red rot resistant cultivar Co 93009 showed induction of four chitinase proteins with molecular masses of 39, 36, 35 and 34 kDa and the intensity of these proteins increased with time from 6 to 42 hr after inoculation. In the susceptible variety CoC 671 an induction of a 35 kDa chitinase protein was recorded and such induction was delayed as compared to the resistant variety where the same induced 24 hr after inoculation in stalk samples.

The mycolytic effect of extracellular enzymes were reported by Viswanathan et al. [44] with the antagonistic microbe *T. harzianum* strain T5, which showed increased levels of activity of N-acetylglucosaminidase and β-1,3-glucanase against *C. falcatum*. Based on these the partial endochitinase cDNA of *Trichoderma harzianum* T5 (246bp) was cloned and sequenced which showed high level of homology with chitinase sequences in the database [91]. The Efficacy of talc-based formulations of fluorescent pseudomonad (FPs) strains (CHAO, EP1 and Pf1) were evaluated as a sett treatment while planting followed by two soil applications in the field by Viswanathan and Samiyappan [92]. It significantly improved vegetative sett germination, improved overall cane growth by 20 to 40%, produced more number of millable canes, ISR activities and less pathogen induced invertase enzyme activity and juice characters *viz.,* sucrose per cent (19.01%). Commercial cane sugar increased by 13.35% as compared to the untreated stalk tissues, after pathogen inoculation. No disease development or less than 1% was recorded in fields treated with *Pseudomonas* sp. compared to 1-2% in control. Ramesh Sundar and Vidhyasekaran [93] reported that a glycoprotein elicitor isolated from the mycelial cell wall of the *C. falcatum* differentially activates the resistant mechanism in suspension cultured cells of sugarcane *viz.,* H_2O_2, lipoxigenase, lipid peroxidation, SOD and catalase in response to early recognition of the pathogen by host cells as compared to the elicitor of *C. lindemuthianum* .

Malathi et al. [94] reported that sugarcane synthesizes a complex mixture of phytoalexins in response to pathogen inoculation in field grown sugarcane varieties *viz.,* Co 93009 and CoC 671, which are resistant and susceptible to red rot respectively. The tissue extracts analyzed by HPLC at different intervals after inoculation revealed the induction of five major compounds in response to pathogen inoculation/injury. Among the five detected compounds only two were found to be induced specifically due to pathogen inoculation *viz.,* luteolinidin and apigeninidin induced specifically in red rot resistant variety in response to attempted pathogen infection and susceptible variety failed to synthesize these compounds.

Peroxidase (PO)

Bradley et al. [95] reported that the increased peroxidase (PO) activity has been correlated with resistance in many plant species including barley, cucurbits, cotton, tobacco and wheat. These enzymes are involved in the polymerization of proteins and lignin or suberin precursor into plant cell wall, thus constructing a physical barrier that could prevent pathogen penetration of cell walls or movement through vessels

Peroxidase is one of the key enzymes involved in the phenyl propanoid pathway and it is involved in the regulation of plant cell elongation, phenol oxidation, polysaccharide cross linking, IAA oxidation, cross linking of extension monomers, oxidation of hydroxy - cinnamyl alcohols into free radical intermediates and wound healing [96].

Singh et al. [97] reported significant increase in PO activity after 4-6 days of pathogen inoculation in red rot moderately resistant cultivar. In sugarcane, the possible involvement of peroxidases in determining red rot resistance has been reported [98 ,99, 100]. Ramesh Sundar et al. [88] treated the, sugarcane cultivar *cv.* CoC 671 with SAR inducers namely Acibenzolar-S-methyl (ASM), salicylic acid (SA) and isonicotinic acid (INA) . Among the treatments, ASM treatment followed by challenge inoculation with *C. falcatum* recorded the maximum level of activity followed by INA and SA. Overall, manifold increase in PO activity was observed due to treatment with SAR inducers as compared to untreated control. Systemic resistance inducers triggered appearance of many low and higher molecular weight isoforms of PO.

The role of peroxidase, catalase and superoxide dismutase enzymes was studied by Asthir et al. [101] inoculated conidia of red rot fungus in two cultivars *viz.,* CoJ 64 (susceptible) and CoS 8436 (resistant) . The resistant cultivar CoS 8436 showed relatively higher activities of peroxidase, catalyses and superoxide dismutase in the intermodal tissues of sugarcane. Later, it was also confirmed by histochemical studies.

Thirupathiraja et al. [102] reported higher rate of peroxidase activity after pathogen inoculation in moderately resistant cultivars *viz.,* BO 91, Co 94008, Co 93009 and Co 86249 followed by moderately susceptible Co 8021 (MS) and highly susceptible cultivars *viz.,* CoC

671 (HS), CoC 92061. Moderately resistant cultivars showed higher enzyme kinetics peak values and the peaks were observed much earlier as compared to moderately susceptible and highly susceptible cultivars. The intensity of isozyme differed in all the cultivars after the pathogen inoculation.

Polyphenol Oxidase (PPO)

Polyphenol oxidase usually accumulates upon wounding in plants. When sugarcane was treated with PGPR (Plant growth promoting Rhizobacteria) before pathogen inoculation, it showed comparatively lesser induction of PPO isoforms than the PGPR untreated plants [103]. In overall, induction of several new PPO isoforms was observed due to treatment with all the three SAR inducers (ASM, INA, SA) as compared to untreated control [88].

Phenylalanine Ammonia - Lyase (PAL)

PAL is the key enzyme in inducing synthesis of salicylic acid (SA) which induces systemic resistance in many plants. Phenylalanine ammonia lyase plays an important role in the biosynthesis of phenolics and phytoalexins [104]. The activation of the phenylpropanoid pathway in plants by environmental stimuli is one of the most universal biochemical stress responses known. Singh et al. [97] reported that, PAL activity increased gradually up to 5 days after pathogen inoculation in moderately red rot resistant variety as compared to susceptible variety and after that it decreased.

Chitinase

Chitinases are PR-proteins which hydrolyze chitin, a major cell wall component 3-10 per cent of higher fungi. The production of chitinases in plants has been suggested to be a part of their defense mechanism against fungal pathogens [105]. Chitinase secreted by microbes are capable of hydrolysing chitin, an insoluble polysaccharide present in the cell wall of higher fungi, insect gut wall and nematodes. These enzymes hydrolyse the chitin present in the cell wall, leading to lysis of the fungal cell [106]. Chitinolytic enzymes inhibit spore germination, germ tube elongation and thus they are thought to be potential bioagents for the suppression of fungal propagules [107].

Viswanathan et al. [44] reported that *Pseudomonas* strain KKM1 treatment directly induced new isoforms of chitinases after pathogen inoculation in the stalk tissues of susceptible variety CoC 671. When it was purified by affinity digestion, the purified chitinases inhibited the mycelial growth of the pathogen.

Electrolyte Leakage Induced by *C. Falcatum* Toxin

Fungal toxin cause serious damage to the cellular functions of host tissue. Toxins are generally products of the pathogen, host or host-pathogens interaction, directly act on living host protoplasm to influence the source of disease development or symptom expression even at very low concentrations. Naik and Vedamurthy [108] and Mohanraj et al. [109] used toxin of *C. falcatum* in the selection of red rot resistant genotypes of sugarcane. Vedamurthy et al. [110] reported that partially purified *C. falcatum* toxin reduced the total uptake of glucose and also inhibited its conversion into insoluble products of cellular metabolism. It also lowered the synthesis of total sugar, which was mainly noticed in callus of susceptible var CoC 671. The red rot pathogen produces a phytotoxin that reproduces some of the symptoms of the disease [111]. This phytotoxin also induces an accumulation of phytoalexins (anthocyanidin pigments) in treated canes similar to that caused by the pathogen [99]. Mohanraj et al. [109] reported that, phytotoxin of *C. falcatum* caused increased electrolyte leakage in susceptible sugarcane varieties and higher levels of phytoalexins (3-deoxyanthocyanidins) in resistant sugarcane varieties. This relationship between phytotoxin induced changes and disease reaction could possibly be used as an additional index to rapidly identify red-rot resistant varieties.

Use of Somaclonal Variation

The conventional breeding method is highly time consuming and takes more than ten years to reach farmer's level. Hence an alternative is to go for *in vitro* techniques [112]. Somaclonal variation has been employed to develop red rot resistant clones in sugarcane [113, 114]. Jalaja et al. [114] was the first to develop resistant variety Co 94012 against red rot by using somaclonal variation from variety CoC 671. The clone was moderately resistant to red rot and smut. Kumar et al. [115] also suggested the possibilities of developing red rot resistant sugarcane variety by employing somaclonal variation . Hence, somaclonal variation opens a new avenue in developing disease resistant varieties of sugarcane.

Conclusion

Despite intensive researches made on the development and management of red rot disease in India, we have not achieved the success up to the desired extent. The management of red rot still remains a mystery. Though several resistant cultivars have been introduced from time to time, the breakdown of resistance and epidemics are still unpredictable. The failure of chemical and biological control may be attributed to the hard rind of the cane and the plant height makes even whorl application difficult. Though a lot of advanced research on identification of the pathogen at molecular levels has been carried out, an easy to use quick identification tool at farm level is lacking. Thus, it seems necessary to carry out more researches in order to develop user friendly techniques to manage red rot in sugarcane. It seems difficult to control red rot or *C. falcatum,* as every entity has a right to live in this world and it is not possible to wipe out any organism or alter the ecological balance created in nature but an effort can be made to minimize the economical losses through management of diseases specially red rot.

REFERENCES

[1] Deerr S (1949). *The history of sugar*. Vol I. Chapman and Hall, London. 258 pp.

[2] Singh O and Waraitch KS (1977). Metabolic changes induced by *Colletotrichum falcatum*. Went. in sugarcane. Sugarcane Pathologist's Newsletter. 19:7-9.

[3] Alexander KC and Viswanathan R (1996). Major diseases affecting sugarcane production in India and recent experiences in quarantine. In *Sugarcane Germplasm Conservation and Exchange*. (Eds. B. J. Croft, C. M. Piggin, E. S. Wallis and D. M. Hogarth), ACIAR Proceedings, Canberra, No 67, 46-48.

[4] Barber CA (1901). Sugarcane diseases in Godavari and Ganjam districts. Madras Dept. *Land Records and Agri. Bull*, 512, 43pp. 181-194.

[5] Agnihothri VP and Singh K (1977). Seed-piece transmissible diseases of sugarcane and their control measures. *Sugar News*, 2: 90-95.

[6] Butler EJ (1906). Fungus diseases of sugarcane in Bengal. *Mem. Dept. Agr. India, Bot.* Ser. 1:2-4.

[7] Butler EJ (1918). Fungi and diseases in plants. An Introduction to the Diseases of Field and Plantation Crops. Thacker, Spink and Co, Calcutta, 547 pp.

[8] Ramakrishnan TS (1941). Studies on the genus *Colletotrichum* II. Physiological studies on *Colletotrichum falcatum* Went. Proc, *Indian Academy of Sciences*. 14:395-411.

[9] Atkinson RS (1938). On the nature of resistance of sugarcane to redroot. *Proceedings of international Society Sugarcane Technologists,* 6, 684-692.

[10] Chona BL, Bajaj BS and Sharma R (1961). A new stoma forming strain of *Colletotrichum falcatum*. Went. *All India Conf. Sugarcane Res & Dev. Workers, Waltair,* (AP) Jan. 1960, 4, 620-622.

[11] Khanna KL (1943). Annual Report, Central Sugarcane Res. Station., Pusa, Bihar, 1940, 40-47.

[12] Rafay SA (1953). Pathogenicity , spread and reversion of *Physalospora tucumanensis*. *Proc. Indian Acad. Sci.,*38(B):99-100.

[13] Duttamajumdar SK (2008). Red rot of sugarcane. *Indian Institute of Sugarcane Research* (IISR), Lucknow, India. Pp.

[14] Viswanathan R (2010). "Plant Disease" Redrot of Sugarcane. *Anmol Publications* Pvt. Ltd, New Delhi pp.304

[15] Rafay SA and Singh VB (1957). A new strain of *Glomerella tucumanensis*. *Current Science,* 26, 19-20.

[16] Agnihotri VP, Budharaja TR and Singh K (1979). Role of diseased setts and soil in the annual recurrence of redrot in sugarcane. *International Sugar Journal* (UK), 82(969) 263-265.

[17] Singh N and Singh GB (1994). Effect of redrot incipient infections in sugarcane seed setts on germination. *Indian Journal of Sugarcane Technology*, 9, 149-151.

[18] Chona BL (1960). Redrot of sugarcane and sugar industry- a review. *Indian Phytopathology,* 33,191-207.

[19] Kumar S, Singh NNP, Kumar V and Dwidevi NB (1994). Deterioration in yield and juice quality parameters in sugarcane due to isolates of redrot pathogen. *Indian Journal of Sugarcane Technology*, 2, 115-122.
[20] Alexander KC and Viswanathan R (2002). Diseases of Sugarcane in India and its rapid diagnosis. In: Sugarcane Crop management, (Eds, S. B. Singh, G. P. Rao, S. Eswaramoorthy), SCITECH publishings LLC, Houton, Texas, USA, pp 10-51.
[21] Viswanathan R, Padmanaban P and Mohanraj D (1997) Growing virulence of redroot pathogen of sugarcane in Tamil Nadu. *Indian Sugar*, 47, 23-30.
[22] Viswanathan R, Ramesh Sundar A, Padmanaban P and Mohanraj D (2002). Redrot disease of sugsrcsne and its management. In: *IPM Systems in Agriculture* vol. 8, *Key Pathogens and Diseases*. (Eds. R. K. Upadhyay, K. G. Mukherji and O. P. Dubey), Aditya Books, New Delhi, pp 277-301.
[23] Agnihotri VP (1983). *Diseases of Sugarcane*. Oxford & IBH Publishing Co., New Delhi, pp. 363.
[24] Kirtikar and Verma HS (1962). A review on effect of sugarcane diseases on yield and juice quality in Uttar Pradesh. *Indian Sugar*, 14,103-108.
[25] Sandhu SS, Bhatti DS and Ratan BK (1969). Extent of losses in sugarcane caused by redrot and (*Physalospora tucumanensis* Speg.) and smut (*Ustilago scitaminae) Syd. Journal of Research*, (PAU, Ludhiana), 6, 341-344.
[26] Satyavar K, Raj A and Virk KS (2002). *Redrot of sugarcane: the research scene in haryana* In: Sugarcane Crop Management,(Eds.s.B. Singh, G.P. Rao, S. Eswaramoorthy), SCITECH publishings LLC, Houton,Texas,USA,pp 109-126.
[27] Islam MN, Rahman MH, Hasan MF and Saud ZA (2002). Post infection changes in nutrient composition and enzyme activity of healthy and redroot affected cane juice. *Pakistan Sugar Journal*, 17, (5) 16-19.
[28] Went FAFC (1893). Het rood snot, *Arch. Java-Suikerindus*, 1, 265-282.
[29] Spegazzini C (1896). Hongas de la cana de azucar. Rev. Fac. Agron. Y Vet. Argentina 2,227-258.
[30] Von Arx JA and Muller E (1954). Die Gattungen der Amerosporen Pyrenomyceten. Beitrage Zur Kryptogamenflora der Schweiz, 11, 5-434.
[31] Butler EJ and Khan HA (1913). Red rot of sugarcane. *Mem, Dept. Agric. India, Bot.* Ser., 6:151-178.
[32] Chona BL and Srivastava DN (1952). The perithicial stage of *Colletotrichum falcatum* Went. In India. *Indian Phytopathology* 5,158-160.
[33] Chona BL and Bajaj BS (1953). Occurrence in nature of *Physalospora tucumanensis* Speg.,the perfect stage of sugarcane redrot organism in India.*Indian Phytopathology* 6, 63-65
[34] Chona BL and Nariani TK (1952). Investigations on the survival of *Colletotrichum falcatum* in soil. *Indian phytopathology* 5,152-157
[35] Singh N and Singh K (1982). Formation of resting structures by *Colletotrichum falcatum* Went. In soil. *Current Science*, 51, 102-104.
[36] Singh N, Lal S and Singh K (1985). Behavior of *Colletotrichum falcatum* under stress condition. *Indian Phytopathology*. 38, 544-545.
[37] Singh, G.P. and Rana, O.S.1968. A new strain of *Glomerella tucumanensis* redrot pathogen in UP. *Indian Sugar*, 18, 537-540.

[38] Gupta, S.C., Singh, M.P and Upadhaya, U.C.1980 . A new biotype of *Colletotrichum falcatum* Went. *Curr. Sci.,* 49:600-601.

[39] Jothi R (1989). Studies on variation in red rot pathogen *Colletotrichum falcatum* Went. On sugarcane, PhD. Thesis, Bharathiyar University, Coimbatore, 254pp.

[40] Duttamajumder SK (2002). A century of redroot disease of sugarcane in India. In: *Sugarcane Crop management,* (Eds. S. B. Singh, G. P. Rao, S. Eswaramoorthy), Sci Tech Publishing LLC, Houston, Texas,USA, pp 52-108.

[41] Duttamajumdar SK, Singh N and Agnihotri VP (1990). Behavior of *Colletotrichum falcatum* under water logged conditions. *Indian Phytopathology*, 43:227-229.

[42] Malathi P, Viswanathan R, Ramesh Sundar A, Prakasam Padmanaban P, Jothi R, Renuka Devi S R and Poongothai M (2010). Variability among *Colletotrichum falcatum* pathotypes used for screening red rot resistance in Sugarcane. *Sugarcane Intl.* 28(2):47-52.

[43] Padmanaban P, Mohanraj D, Viswanathan R, Madhusudhanrao M, Prakasam N, Jothi R and Alexander KC (1996). Differential interaction of sugarcane clones to pathotypes of *Colletotrichum falcatum* Went. *Sugarcane* 4:16-20.

[44] Viswanathan R, Nandakumar R and Samiyappan R (2003). Role of pathogenesis-related proteins in rhizobacteria-mediated induced systemic resistance against *Colletotrichum falcatum* in sugarcane. *J. Plant Dis. Protect*. 110(6):524-534.

[45] Singh RP and Lal S (1999). Red Rot. Pages153-158 in: A Guide to Sugarcane Diseases. CIRAD-ISSCT publications. ISBN 2-87614-386-0.

[46] Singh RP, Lal S and Singh K (1988). Effect of ambient temperature on red-rot development in sugarcane. *Indian Phytopath*. 41:86-91

[47] Alexander KC, Rao M M and Mohanraj D (1985). Disease reaction catalogue on genetic resources II. Sugarcane Breeding Institute, Coimbatore, Tamil Nadu, India. p.226.

[48] Kumar A and Virk KS (2001). Pathogenic variability in *Colletotrichum falcatum* in Haryana. *Indian Phytopath*. 54:505.

[49] Nageswararao GV and Achutaramarao M (2004). Occurrence of a new virulent pathotype of *Colletotrichum falcatum* on sugar cane in Andhra Pradesh. *J. Mycol.Plant* Pathol. 34:119-121.

[50] Nageswararao GV and Patro T S S K (2005). Pathotypes in *Colletotrichum falcatum* Went. and identification of resistant sugarcane clones. *J. Mycol. Plant Pathol*. 35: 305-309.

[51] Satyavir (2003). Red rot of sugarcane - Current scenario. Indian Phytopath. 56:245-254.

[52] Satyavir, Singh N, Virk KS, Nageswarrao GV, Singh H and Mishra SR (2001). Pathogenic variability in sugarcane red rot system. Pages 109-114 in: Proc. National Symp. Role of Resistance in Intensive Agriculture. S. Nagarajan, and O.P. Singh, eds. Kalyani Publishers, Ludhiana.

[53] Malathi P, Viswanathan R and Jothi R (2006). Specific adaptation of *Colletotrichum falcatum* pathotypes to sugarcane cultivars. *Sugar Tech* 8(1):54-[54].

[54] Mohanraj D, Kumaresan S and Sreenivasan TV (2002a). Molecular characterization of isolates of the sugarcane red rot pathogen. *Indian Phytopath.* 55:147-151.

[55] Sangeetha Panicker (2011a). Identification of suitable disease progress model for red rot of sugarcane caused by *Colletotrichum falcatum*. Proceedings of the International Sugar Conference held at New Delhi during 21st to 25th November, 2011.

[56] Viswanathan R and Samiyappan R (2002). Induced systemic resistance by fluorescent pseudomonads against red rot disease of sugarcane caused by *Colletotrichum falcatum.* Crop Prot. 21:1-10.

[57] Sangeetha Panicker (2011b). "Sugarcane Doctor"- an interactive software for the identification and management of sugarcane diseases. *Proceedings of the International Sugar Conference held at New Delhi during 21st to 25th November*, 2011.

[58] Singh SB, Rao GP and Easwaramoorthy S (2002). Sugarcane Crop Management, SCI TECH Publishing LLC, Texas, USA. p.334.

[59] Viswanathan R, Samiyappan R and Padmanaban P (1998). Specific detection of *Colletotrichum falcatum* in sugarcane by serological techniques. *Sugarcane* 3:18-23.

[60] Hiremath L and Naik GR (2004). Rapid diagnosis of sugarcane red rot by Dot-immunobinding assay (DIBA) technique. *Ind. J. Biotechnol.* 3(4):542-545.

[61] Khalid AIM Bukhari, Nithya K, Valluvaparidasan V, Paranidharan V and Velazhahan R (2012). Detection of *Colletotrichum falcatum* causing redrot of sugarcane by enzyme linked immunosorbent assay. *Archives of Phytopathology and Plant Protection.*45:823-830

[62] Lingayya Hiremath and Naik GR (2004). Rapid diagnosis of Sugarcane redrot by Dot immunobinding assay (DIBA) technique. *Indian Journal of biotechnology.* 3: 542-545.

[63] Nithya K, Khalid AIMBukhari, Valluvaparidasan V, Paranidharan V and Velazhahan R (2012). Molecular detection of *Colletotrichum falcatum* causing redrot disease of sugarcane using SCAR marker. *Ann. Appl. Biol.* 160:168-173.

[64] Narender Kumar, Jhang T, Satyavir and Sharma TR (2010). Molecular and Pathological characterization of *Colletotrichum falcatum* infecting subtropical Indian Sugarcane. *J. Phytopathol* 159 : 260-267

[65] Saksena P, Vishwakarma SK, Tiwari AK, Singh A and Kumar A (2013).

[66] Pathological and Molecular variation in *Colletotrichum falcatum* isolates causing redrot of sugarcane in North Western zone of India. *Journal of Plant Protection Research.* 53: 37.

[67] Samuels GJ and Siefert KA (1995). The impact of molecular characters on systematics of filamentous ascomycetes. *Annu. Rev. Phytopathol.* 33:37-67.

[68] Correll JC, Rhoads DD and Guerber J C (1993). Examination of mitochondrial DNA restriction fragment length polymorphisms, DNA fingerprints and randomly amplified polymorphic DNA of *Colletotrichum orbiculare. Phytopathology* 83:1199-1204.

[69] Fabre JV, Julien J, Parisot D and Dron M (1995). Analysis of diverse solates of *Colletotrichum lindemuthianum* infecting common bean using molecular markers. *Mycol. Res.* 99:429-435

[70] Williams JG K, Kubelik AR, Rafalski LA and Tingey SV (1990). DNA polymorphisms amplified by arbitrary primers are useful as genetic markers. *Nucleic Acids Res.* 18:7213-7218.

[71] Freeman S and Rodriguez RJ (1995). Differentiation of *Colletotrichum* species responsible for anthracnose of strawberry by arbitrarily primed PCR. *Mycol. Res.* 99: 501-504.

[72] Vos P, Hogers R, Bleeker M, Reijins M, Van de Lee T, Hornes M, Frijters A, Pot J, Peleman J, Kuiper M and Zabeau M (1995). AFLP: a new technique for DNA fingerprinting. *Nucleic Acids Res.* 23:4407-4414.

[73] Freeman S and Katan T (1997). Identification of *Colletotrichum* species responsible for anthracnose and root necrosis of strawberry in Israel. *Phytopathology* 87:516-521

[74] White TJ, Bruns T, Lee S and Taylor J (1990). Amplification and direct sequencing of fungal ribosomal RNA genes for phylogenetics. Pages 282-287 in: PCR Protocols: A Guide to Methods and Applications. M. A. Innis, , D. H. Gelfand, J. J. Sninsky, and T. J. White, eds. Academic Press, San Diego, CA, USA.

[75] Sreenivasaprasad S, Mills P, Meehan BM and Brown A (1996). Phylogeny and systematics of 18 *Colletotrichum* species based on ribosomal DNA spacer sequences. *Genome 39*:499-512.

[76] Madan VK, Mandal B, Ansari MI, Srivastava A, Soni N, Solomon S, Agnihotri VP, Mandal B and Srivastava A (2000). RAPD-PCR analysis of molecular variability in the red rot pathogen (*Colletotrichum falcatum*) of sugarcane. *Sugarcane Intl.* 3:5-8.

[77] Suman A, Sunitha Lal, Shasany AK, Gaur A and Singh P (2005). Molecular assessment of diversity among pathotypes of *Colletotrichum falcatum* prevalent in subtropical Indian sugarcane. *World J. Microbiol. Biotechnol.* 21:1135-1140.

[78] Alvi AK, Iqbal J, Shah AH and Pan YB (2008). DNA based genetic variation for red rot resistance in sugarcane. *Pak. J. Bot.* 40:1419-1425.

[79] Mishra M K and Behera B (2009). Pathogenic and molecular variability of *Colletotrichum falcatum* Went. Isolates from sugarcane with red rot disease symptoms. *J. Crop Sci. Biotechnol.* 12(1):31-36.

[80] Kumar N, Tripta J, Satyavir and Sharma T R (2011). Molecular and pathological characterization of *Colletotrichum falcatum* infecting subtropical Indian sugarcane. *J. Phytopathology* 159:260-267.

[81] Wijesekara HTR, Agarwal R and Agarwal DK (2005). Morphological and molecular characterization of five *Colletotrichum* species from India. *Indian Phytopath.* 58:448-453.

[82] Hammerschmidt R and Kuc J (1995). *Induced systemic resistance to disease in plants.* Kluwer Academic Publishers, Dordrecht, the Netherlands. p.182.

[83] Van Loon LC, Bakker PAHM and Pieterse CM J (1998). Systemic resistance induced by rhizosphere bacteria. Annu. Rev. Phytopathol. 36:453-483.

[84] Vidhyasekaran P and Muthamilan M (1999). Evaluation of powder formulation of *Pseudomonas fluorescens* Pf1 for control of rice sheath blight. *Biocontrol Sci. Technol.* 9:67-74.

[85] Viswanathan R (1999). Induction of systemic resistance against red rot disease in sugarcane by plant growth promoting rhizobacteria. *Ph.D. Thesis,* TNAU, Coimbatore, India. p.175 .

[86] Viswanathan R and Samiyappan R (1999). Induction of systemic resistance by plant growth promoting rhizobacteria in sugarcane against red rot disease. *Sugar Tech* 1: 67-76.

[87] Ramesh Sundar A, Viswanathan R and Nagarathinam S (2009). Induction of systemic acquired resistance (SAR) using synthetic signal molecules against *Colletotrichum falcatum* in sugarcane. *Sugar Tech:* 11(3):274-281.

[88] Ramesh Sundar A, Velazhahan R, Viswanathan R, Padmanaban P and Vidhyasekaran P (2001). Induction of Systemic Resistance to *Colletotrichum falcatum* in Sugarcane by a synthetic signal molecule, Acibenzolar-S-Methyl (CGA-245704), *Phytoparasitica* 29(3):231-242.

[89] Ramesh Sundar A, Viswanathan R, Malathi P and Padmanaban P (2006). Mechanism of resistance induced by plant activators against *Colletotrichum falcatum* in sugarcane, Arch. Phytopathol. *Plant Protect*. 39(4):259 - 272.

[90] Viswanathan R and Samiyappan R (2001). Role of chitinases in *Pseudomonas* spp. induced systemic resistance against *Colletotrichum falcatum* in sugarcane. *Indian Phytopath*. 54:418-423.

[91] Viswanathan R, Nandakumar R and Samiyappan R (2003). Involvement of pathogenesis-related proteins in *Pseudomonas* spp mediated induced systemic resistance against *Colletotrichum falcatum* in sugarcane. *Journal of Plant Diseases and Protection* 110,:524-534.

[92] Viswanatha R, Malathi P, Ramesh Sundar A, Aarthi S, Premkumari SM and Padmanaban P (2005). Differential induction of chitinase and thaumatin- like proteins in sugarcane in response to infection by *Colletotrichum falcatum* causing redrot disease. *J. Plant Dis. Protect*. 112:417-425.

[93] Viswanathan R and Samiyappan R (2002). Induced systemic resistance by fluorescent pseudomonads against red rot disease of sugarcane caused by *Colletotrichum falcatum. Crop Prot*. 21:1-10.

[94] Ramesh Sundar A and Vidhyasekaran P (2003). Induction of defence related biochemical changes by elicitors of red rot pathogen and a non pathogen in sugarcane cell culture. *Indian Phytopath*. 56(3):255-261.

[95] Malathi P, Padmanaban P, Viswanathan R and Mohanraj D (2008). Interaction between *Colletotrichum falcatum* pathotypes and bio control agents. *Arch. Phytopath. Plant Prot*. 41:311-317.

[96] Bradley DJ, Kjellborn P and Lamb C (1992). Elicitor and wound induced oxidative cross-linking of a plant cell wall proline-rich protein: A novel, rapid defence response. Cell 70:21-30.

[97] Vidhyasekaran P, Rabindran R, Muthamilan M, Nayar K, Rajappan K, Subramanian N and Vasumathi K (1997). Development of powder formulation of *Pseudomonas fluorescens* for control of rice blast. *Plant Pathol*. 46:291-297

[98] Singh J, Chandra P, Rao GP and Singh HN (1993). Biochemical studies on resistance to red rot in sugarcane. *Sugarcane* 6.

[99] Viswanathan R Mohanraj D and Padmanaban P (1996a). Role of red rot pigments in relation to red rot resistance in sugarcane. *Indian J. Sugarcane Technol*. 11:151-154.

[100] Viswanathan R, Mohanraj D, Padmanaban P and Alexander KC (1996b). Synthesis of phytoalexins in sugarcane in response to infection by *Colletotrichum falcatum* Went. Acta Phytopathol. *Entomol. Hung*. 31:229-237.

[101] Ramesh Sundar A, Viswanathan R, Mohanraj D and Padamanaban P (1999). Studies on the possible use of pathogen toxin as a marker for redrot resistance in sugarcane. Acta Phytopathol. *Entomol. Hung*. 34:211-217.

[102] Asthir B, Preet K, Batta SK and Sharma B (2009). Role of antioxidative enzymes in red rot resistance in sugarcane. *Sugar Tech* 11(3):282-287.

[103] Thirupathiraja C, Viswanathan R and Padmanaban P (2004). Time course of peroxidase accumulation in sugarcane cultivars in response to *Colletotrichum falcatum* infection. Sugar Tech 6(1&2):47 -52.

[104] Viswanathan R and Samiyappan R (1999). Induction of systemic resistance by plant growth promoting rhizobacteria in sugarcane against red rot disease. *Sugar Tech* 1: 67-76.

[105] Daayf F, Bel-Rhlid R and Belanger RR (1997). Methyl ester of *p*-coumaric acid: A phytoalexin-like compound from long English cucumber leaves. *J. Chem. Ecol.* 23:1517-1526.

[106] Schlumbnaum A, Mauch F, Vogeli WR and Boller T (1986). Plant chitinases are potent inhibitors of fungal growth. *Nature* 324:365-367.

[107] Kramer KJ and Muthukrishnan S (1997). Insect chitinase: molecular biology and potential use as biopesticides. Insect Biochem. *Mol. Biol.* 27:887-900.

[108] Frankowski J, Lorito M, Scala F, Schmid R, Berg G and Bahl H (2001). Purification and properties of two chitinolytic enzymes of *Serratia plymuthica* HRO-C48. *Arch. Microbiol.* 176:421-426.

[109] Naik GR and Vedamurthy AB (1997). *In vitro* evaluation of red rot toxin influence on sugarcane (*Saccharum officinarum* L) var. CoC *671. Curr. Sci.* 73(4):367-369.

[110] Mohanraj D, Padmanaban P and Karunakaran M (2003). Effect of phytotoxin of *Colletotrichum falcatum* Went. (*Physalospora tucumanensis*) on sugarcane in tissue culture. Acta Phytopathol. *Entomol. Hung*. 38:21-28.

[111] Vedamurthy AB, Sateesh MK and Naik GR (2007). *In vitro* effect of red rot toxin on carbohydrate metabolism of sugarcane var. CoC 671. *Indian Sugar* 57(9):43-48.

[112] Mohanraj D, Padmanaban P and Karunakaran M (2002b). Association of phytotoxin produced by *Colletotrichum falcatum* Went in the red rot disease of sugarcane. Sugarcane 21-23.

[113] Tiwari AK, Bharti YP, Mishra N, Tripathi S, Lal M, Sharma PK, Rao GP and Sharma ML (2010). Biotechnological approaches for improving sugarcane crop with special reference to disease resistance. *Acta Phytopathologica et Entomologica Hungrica* 45: 235–249.

[114] Sreenivasan TV and Jalaja NC (1995). Utility of tissue culture technology in sugarcane improvement. In: *Application of Bio-technology for sustainable sugarcane production.* Ann Conv DSTA, Pune 44: 1–9.

[115] Jalaja NC, Sreenivasan TV, Pawar SM, Bhoi PG and Garker RM (2006). Co 94012—a new sugarcane variety through somaclonal variation. *Sugar Tech* 8: 132–136.

[116] Pankaj Kumar, Agarwal A, Tiwari AK, Lal M and Jabri M R A (2012). Possibilities of Development of Red Rot Resistance in Sugarcane Through Somaclonal ariation. *Sugar Tech*: 192-194.

About the Editors

Dr Ajay K Tiwari is working as Principal Investigator in Department of Science & Technology, New Delhi, India funded project. He did his Ph D in 2011 on Cucurbit viruses from Biotechnology department of CCS University, Meerut, UP, India. Dr Tiwari is regular member of British Society of Plant Pathology, Indian Phytopathological Society, Sugarcane Technologists Association of India, International Society of Sugarcane Technologists, Society of Sugarcane Research and Promotion, Prof H S Srivastava foundation, Society of Plant Research. He has published 38 research article, 5 review article in national and International journals, 3 book chapters in edited books and also submitted more than 90 nucleotide sequences of plant pathogens in Genbank to his credit. He is regular reviewer of many International journals and also member of editorial board. He has been awarded Young Researcher award in Italy 2011, Young Scientist award by DST-SERB and also nominated for Narshiman Award by Indian Phytopathological Society, India. Dr Tiwari is also recipient of many International travel award given by DST, DBT, CSIR from India, PATHOLUX from Luxembarg and IOM from Brazil. He has visited China, Italy, Germany for conferences/workshop. He has been involved in the research on molecular characterization and management of agricultural plant pathogens for the last 9 years. Currently he is working on Molecular characterization of sugarcane phytoplasma and their secondary spread in nature.

Dr. Madan Lal is presently working as a Scientific Officer and Head, Tissue Culture Division, U.P. Council of Sugarcane Research, Shahjahanpur, India. Dr Lal has an experience of over 28 years of research in the fields of plant morphogenesis and tissue culture. After post graduation in Botany in 1982, he worked for his Ph.D. with Professor (Late) V.S. Jaiswal, Banaras Hindu University, Varanasi. He has made several contributions on *in vitro* culture of sugarcane. Rapid multiplication of newly released varieties of sugarcane through *in vitro* micropropagation and sugarcane improvement through somaclonal variation has been his main fields of interest. With a view to reduce the cost of micropropagated plants, he has developed low cost protocols for efficient micropropagation of several varieties of sugarcane. He has also worked on some biochemical and molecular changes taking place during *in vitro* morphogenesis in several plants. He has successfully encapsulated the somatic embryos and young meristems of sugarcane and achieved germination of artificial seeds *in vitro*. Numerous field demonstrations using micropropagated plantlets of sugarcane had been conducted by him at the farms of UPCSR, cane growers and sugar factories of Uttar Pradesh and Bihar to motivate them for adopting the micropropagated seed material of sugarcane. Dr. Lal has published over 70 research papers in Indian and International journals. He has also published a book entitled 'Tissue Culture Based Sugarcane Farming' to his credit.

Dr. Ajay Kumar Singh is presently working as Senior Scientist (Agronomy), Division of Transfer of Technology, Indian Institute of Natural Resins and Gums (ICAR), Ranchi (Jharkhand). After obtaining his master degree in agriculture (M.Sc. Ag., Agronomy) in 2003 from G.B. Pant University of Agriculture and Technology, Pantnagar (Uttarakhand), he earned his Ph.D. degree in Agronomy from the same University in 2007. He joined U.P. Council of Sugarcane Research, Shahjahanpur as Scientific Officer (Agronomy) in 2001 and later as Senior Scientific Officer and Head of Agronomy Division. In the year 2012, he joined as Senior Scientist (Agronomy) in Ranchi, ICAR. He has a vast experience in the fields of sugarcane research, extension and development. He has published more than 20 research papers in different journals of national and international repute and contributed several articles in the books.

INDEX

A

B

C

F

G

H

I

J

K

L

M

N

O

P

Q

R

S

T

U

V

W

Y

Z